# Advances in
# Ceramic Armor II

T0181768

# Advances in Ceramic Armor II

*A Collection of Papers Presented at the 30th International Conference on Advanced Ceramics and Composites January 22–27, 2006, Cocoa Beach, Florida*

Editor

Lisa Prokurat Franks

General Editors

Andrew Wereszczak

Edgar Lara-Curzio

**A JOHN WILEY & SONS, INC., PUBLICATION**

Published by John Wiley & Sons, Inc., Hoboken, New Jersey
Published simultaneously in Canada.

For general information on our other products and services please contact our Customer Care
Department within the U.S. at 877-762-2974, outside the U.S. at 317-572-3993 or fax 317-572-4002.

Wiley also publishes its books in a variety of electronic formats. Some content that appears in print,
however, may not be available in electronic format.

*Library of Congress Cataloging-in-Publication Data is available.*

ISBN-13 978-0-470-08057-3
ISBN-10 0-470-08057-4

10 9 8 7 6 5 4 3 2 1

# Contents

## Glass and Transparent Ceramics

## Other Opaque Ceramics

## Damage and Testing

# Preface

For the fourth year in a row, modelers, experimentalists, processors, testers, fabricators, manufacturers, managers, and ceramists gathered to present and discuss the latest issues in ceramic armor during the 30th International Conference and Exposition on Advanced Ceramics & Composites, January 22–27, 2006 in Cocoa Beach, Florida. This annual meeting is organized and sponsored by the American Ceramic Society (ACerS) and the ACerS Engineering Ceramics Division.

For each of the past four years, we have been privileged to have technically proficient, enthusiastic, and lively presentations and discussions, and this meeting continued the tradition on a truly international scale. The U.K., Japan, Italy, India, Germany, and China were represented in this unclassified forum.

On the first day of the session, Mr. William A. Gooch, Jr. of the Army Research Laboratory and Mr. Roger W. Engelbart of Boeing-Phantom Works presented invited talks entitled, "Overview of the Development of Ceramic Armor Technology: Past, Present, and the Future" and "The Role of Nondestructive Evaluation (NDE) in Armor Development", respectively.

The first paper in this proceedings is, "A Review of Computational Ceramic Armor Modeling" by Dr. Charles E. Anderson Jr. of Southwest Research Institute; a contribution from his invited talk given at the ceramic armor session in 2005. The remaining 20 papers were presented in 2006 and are generally grouped by ceramic material with silicon carbide and transparent materials making up the majority and leading off this section; the presence and role of different damage modes in ceramics is a consideration in a number of papers and damage and testing are primarily addressed in the concluding papers of this section.

Opaque ceramics have been the primary focus of ceramic research in most years past, but with the advent of the conflict in Iraq and the prevalence of tactical and security vehicles in the conflict, transparent materials have gained much attention. In this short preface it is impossible to discuss each excellent contribution that follows, but for the reader it should be noted that although we have outstanding efforts in ceramics for armor the "figure of merit" for armor material design, or the "keystone" laboratory bench test to predict ballistic performance with certainty, continues to elude ballistic material designers and experimentalists.

Special thanks must go to Dr. James W. McCauley, Senior Research Engineer (ST) at the Army Research Laboratory and past President of ACerS, who keeps ceramic armor fresh for both new and seasoned investigators, and ensures much needed support from the academic, government, and industrial communities. Kudos must go to the Organizing Committee for ceramic armor and their Government agencies; they lend their time, talents, name, and resources, which have helped us grow over these last four years from a focus session to a recognized symposia in 2006.

It was a privilege to work with my co-organizer, Dr. Jeffrey J. Swab of the Army Research Laboratory, and for the 2007–08 academic year, the United States Military Academy at West Point, New York. Dr. Swab's genuine scientific curiosity and integrity, proficient management skills, enthusiasm for excellence, and focus on our customer, the Soldier, have solidified my commitment to continue with him as a co-organizer for the 2007 Ceramic Armor symposium.

LISA PROKURAT FRANKS

# Introduction

This book is one of seven issues that comprise Volume 27 of the Ceramic Engineering & Science Proceedings (CESP). This volume contains manuscripts that were presented at the 30th International Conference on Advanced Ceramic and Composites (ICACC) held in Cocoa Beach, Florida January 22–27, 2006. This meeting, which has become the premier international forum for the dissemination of information pertaining to the processing, properties and behavior of structural and multifunctional ceramics and composites, emerging ceramic technologies and applications of engineering ceramics, was organized by the Engineering Ceramics Division (ECD) of The American Ceramic Society (ACerS) in collaboration with ACerS Nuclear and Environmental Technology Division (NETD).

The 30th ICACC attracted more than 900 scientists and engineers from 27 countries and was organized into the following seven symposia:

- Mechanical Properties and Performance of Engineering Ceramics and Composites
- Advanced Ceramic Coatings for Structural, Environmental and Functional Applications
- 3rd International Symposium for Solid Oxide Fuel Cells
- Ceramics in Nuclear and Alternative Energy Applications
- Bioceramics and Biocomposites
- Topics in Ceramic Armor
- Synthesis and Processing of Nanostructured Materials

The organization of the Cocoa Beach meeting and the publication of these proceedings were possible thanks to the tireless dedication of many ECD and NETD volunteers and the professional staff of The American Ceramic Society.

ANDREW A. WERESZCZAK
EDGAR LARA-CURZIO
General Editors

*Oak Ridge, TN (July 2006)*

# A REVIEW OF COMPUTATIONAL CERAMIC ARMOR MODELING

Charles E. Anderson, Jr.
Southwest Research Institute
P.O. Drawer 28510
San Antonio, TX 78228-0510

## ABSTRACT

Computational ceramic armor models are reviewed, with an emphasis on a historical perspective. Some discussion concerning the current state of the art is provided, including a summary of issues concerning the strength of *in-situ* comminuted ceramic.

## INTRODUCTION

Focused sessions on armor ceramics became part of the *International Conference & Exposition on Advanced Ceramics & Composites* annual meetings in 2003. Each year, the Organizers of these focused sessions have selected a topic area and presenter for the first (keynote) talk to start the conference, with the objective of providing an historic overview of some aspect of armor ceramics. Computational ceramics armor modeling was the focused topic for the 2005 meeting. This article documents the presentation that was made at that meeting. The article does not provide an exhaustive review of all the work done in this area, but it does highlight significant progress and difficulties in computational ceramic modeling.

## THE LIGHT ARMOR PROGRAM

Mark Wilkins developed the first computational ceramic armor model, circa 1968, as part of the multi-year light armor program, funded by DARPA, to defeat rifle-fired, armor-piercing (AP) bullets [1-5]. Weight is always an issue with armor, so materials are pushed to their limit, that is, failure. As Wilkins stated: "The application of materials to light armor is unusual because material properties are utilized in the region of material failure, i.e., if the armor doesn't fail for a given ballistic threat, it could be made lighter" [3]. Numerical simulations were an integral portion of the light armor program, providing illumination and guidance.

Early on in the light armor program, Wilkins realized that a lightweight system had requirements for different, and conflicting, material properties. A hard element is needed to erode and decelerate the bullet. A ductile element is required to capture the remnants of the eroded bullet. Thus, materials with different properties need to be assembled in the most advantageous way. Photographs of an experiment, Fig. 1, conducted at Southwest Research Institute (SwRI) depict the response of an AP bullet against a $B_4C$ ceramic tile glued to an aluminum (6061-T6) substrate. The front view shows the damage to the ceramic, and the side view shows the deformation of the aluminum substrate plate. Horizontal lines were drawn on the back of the substrate plate to assist in visualizing the deformation. As can be seen, the substrate plate absorbs some of the kinetic energy through deformation.

Wilkins conducted simulations into metallic targets [1] prior to investigating ceramics. The threat bullet was the 7.62-mm armor-piercing APM2. However, Wilkins developed a monolithic 0.30-cal bullet as a surrogate projectile for the APM2 bullet, largely to decrease the scatter in experimental data resulting from fracturing of the hard steel core in the APM2 bullet. Muzzle velocity for the bullets is 820-850 m/s. The physical characteristics of these two bullets

1

are summarized in Table I, and schematics of the bullets are shown in Fig. 2. The surrogate bullet has a penetration performance similar to that for the APM2 bullet into hard targets.

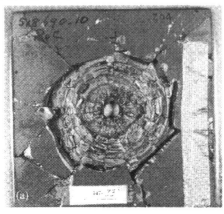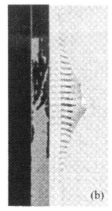

Fig. 1. Post-test photograph of impact of AP bullet against ceramic/aluminum target: (a) front view of ceramic element; (b) side view of target showing deformation of aluminum element.

Table I. Physical Properties of the 7.62-mm AP Bullets

| 7.62-mm APM2 Bullet | 7.62-mm Surrogate AP Bullet |
| --- | --- |
| Mass: 10.74 g | Mass: 8.32 g |
| Length: 3.53 cm | Length: 2.81 cm |
| Core Mass: 5.25 g | Nose: 55° cone |
| Core Length: 2.74 cm | Hardness: $R_c$ 55 |
| Core Hardness: $R_c$ 62-65 | |

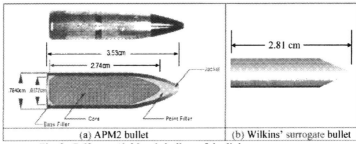

(a) APM2 bullet | (b) Wilkins' surrogate bullet

Fig. 2. 7.62-mm (0.30-cal) bullets of the light armor program.

This activity took place during the days of (Hollerith) computer cards, one of which is shown in Fig. 3. Holes, representing a line of a FORTRAN code were punched into these cards; "fingers" read the punched holes when fed into a card reader. Two thousand cards came in a box, and a large program like HEMP required 3 to 5 boxes for the entire program. (It is noted that it was a "bad" day when a box of cards was dropped. and a *very* bad day if the cards weren't sequenced in columns 73-80).

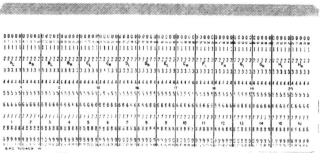

Fig. 3. 80-column Hollerith computer card.

Constraints on Modeling

Before proceeding, it is useful to describe constraints that are imposed on a model if the model is to assist in understanding and/or design. It must be demonstrated that the model captures essential features of observed phenomena, and it must be demonstrated that the model provides reasonable agreement with experimental data. Otherwise, there can be no confidence in any predictions of the model. Further, model parameters should be connected to physical parameters (material properties), and if possible, these physical parameters should be determined from independent laboratory experiments. As will be seen, however, it is sometimes necessary to conduct parametric simulations and determine a value for a model parameter through matching to ballistic experiments. Finally, the same set of parameters should be applicable to a variety of types of experiments.

The above constraints may seem obvious; however, the field is littered with assertions of modeling successes, but only because model parameters were adjusted to replicate an experimental results. This is "self-consistency." Self-consistency is a necessary, but not a sufficient, condition for the validity of a model. This is why, for ballistics impact, it must be demonstrated that the same set of parameters is applicable over a range of impact velocities.

Wilkins' Ceramics Model

The ceramics model developed by Wilkins is applicable to thin (on the order of a projectile diameter) ceramic tiles. The model was implemented in the two-dimensional finite difference code HEMP [6-7]. A primary emphasis of the model was to simulate the development of the fracture conoid, which was observed in experiments. It took until Model 17 before Wilkins was satisfied that he had something that represented the behavior of a ceramic tile to impact [8].[*]

The Wilkins' ceramic model is a tensile failure model. When the maximum principal stress of a cell exceeds a tensile stress criterion ($\sigma > \sigma_f$ ), where the stress is positive in tension, fracture is initiated within the computational cell (see Fig. 4). But additional criteria had to be applied, otherwise all the zones tended to fail within a few computational cycles. An internal

---

[*] I do not remember whether it was really Model #17 or another (fairly large) number. The important point is that it took a while before Wilkins felt that simulations using the model replicated experimental reality.

state variable $\varphi_f$ is used to track damage within a computational cell. Once initiated, fracture is assumed to propagate through the cell at some fraction of the shear wave speed, specified by the parameter $f_1$ (see Fig. 5). Also, once fracture is initiated, the fracture continues within the cell until complete, that is, $\varphi_f$ starts at 0 and fracture continues until $\varphi_f = 1$. As the damage propagates, the cell is progressively weakened, as given in Fig. 5. Thus, it takes a finite amount of time (the time it would take a crack—moving at some fraction of a shear wave—to propagate across a characteristic dimension of the computational cell) before the cell has completely fractured.

An additional fracture initiation criterion also had to be implemented: fracture initiation of a computational cell could only occur at surfaces (including material interfaces) or if a neighboring (immediately adjacent) cell had completely fractured, $\varphi_f = 1$, subject to the criterion that $\sigma > \sigma_f$ at the specified computational time step. In this manner, damage propagated on the computational grid like a crack, at a fraction ($f_1$) of the shear wave velocity. It is important to note that damage (fracture) will arrest if the tensile stress is less than the tensile stress fracture criterion $\sigma_f$.

---

- Fracture initiates on surfaces where the maximal principal stress is greater than $\sigma_f$: $\sigma > \sigma_f$ (stress is positive in tension)
- A fracture may also initiate within a cell if the maximal principal stress is greater than $\sigma_f$ ($\sigma > \sigma_f$) and a neighboring cell has completely fractured, i.e., $\varphi_{f_{neighbor}} = 1$

---

Fig. 4. Wilkins' ceramic model: fracture initiation criteria.

---

- The ceramic material has a parameter $\varphi_f$: $0 \leq \varphi_f \leq 1$

  $\varphi_f = 0$    no fracture has occurred

  $\varphi_f = 1$ material in cell is completely fractured

- If a cell has fractured, it continues to fracture until fracture is complete

  $\varphi_f^{n+1} = \varphi_f^n + \Delta\varphi_f$ until $\varphi_f^{n+1} = 1$

  where

  $$\Delta\varphi_f = f_1 \frac{C_{shear} \cdot \Delta t^n}{X} \qquad 0 < f_1 \leq 1$$

  $C_{shear} = \sqrt{G/\rho} \equiv$ shear wave speed

  $X =$ characteristic length of cell $\left(e.g., X = \sqrt{(\Delta x)(\Delta y)}\right)$

- Cell progressively softens during fracture:

  $Y = (1 - \varphi_f)Y_{intact}$

---

Fig. 5. Wilkins' ceramic model: fracture propagation with a cell and cell strength.

The Wilkins' ceramic model was implemented into an early version of CTH [9-10]. To examine some essential results using the Wilkins' model, a simulation of the 7.62-mm AP simulant impacting a 7.62-mm-thick B₄C tile glued to a 6.35-mm-thick 6061-T6 aluminum substrate was conducted. The ballistic limit for this target is 820 m/s [3]. The results for early time steps are shown in Fig. 6. A fracture conoid begins within the first few microseconds after impact. At approximately 5 μs, an axial crack is initiated at the ceramic-aluminum interface. The fracture conoid proceeds toward the substrate and effectively limits the amount of ceramic that participates in the impact process. The axial crack moves back toward the impact surface, and by 10 μs, the ceramic underneath the penetrator has completely failed.

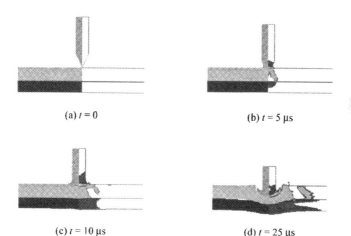

(a) $t = 0$                    (b) $t = 5$ μs

(c) $t = 10$ μs                (d) $t = 25$ μs

Fig. 6. Simulations results for the impact of an AP projectile into a B₄C/Al target using the Wilkins' ceramic model.

The computational results are very zone-sized dependent because of the stress-dependent initiation criterion. Wilkins used 40 zones across the radius of the projectile [8]. Even with today's computational power, 40 zones across the radius of the projectile is a very large number (5 to 10 zones across the radius are typical numbers for many 2-D simulations). Circa 1968, 40 zones to resolve the radius was a phenomenally large number. To put this in perspective, with 40 zones across the radius of the bullet, then there were approximately 17,700 computational cells in the problem (assuming nominally square zoning). The time step is proportional to the smallest zone dimension divided by the material sound speed, and since the sound speed of ceramics is very high (~10 km/s for B₄C), this gives a maximum time step of 0.01 μs. The "grind time" (the amount of computer time per zone per time step) for the supercomputers at that time was on the order of 1 ms per zone cycle. Since HEMP is a Lagrangian code, zonal distortion limited the time step even further. Wilkins carried his calculations out for ~1400 cycles (~6 μs), which took about 6-7 hours CPU time to reach 6 μs; wall clock time was probably significantly longer.

MODIFIED WILKINS' CERAMIC MODEL

In 1991, Walker and Anderson [9] placed the Wilkins' model into CTH [10] and reproduced Wilkins' results. Starting about 1996, the model was used extensively in the DARPA/Army ultra-lightweight body armor program. The failed ceramic material in the original Wilkins' model did not have strength. (Wilkins surely added this later, but the strength of the failed material is not mentioned in the description of the model that is summarized in his reports [3-4].) We found that the failed material had to have some sort of strength; else, the bullet easily perforated the target with a significant amount of its original kinetic energy [11]. Thus, the first modification to the Wilkins' ceramic model was to model the failed ceramic material with a Drucker-Prager model, characterized by a slope $\beta$ and a cap $\overline{Y}$. When fracture was initiated within a computational cell (applying the same initiation and propagation criteria as specified in Figs. 1 and 2), the cell strength went from that of an intact material to failed material, as summarized in Fig. 7, as $\varphi_f$ transitioned from 0 to 1.

$$Y = \left(1 - \varphi_f\right)Y_{\text{intact}} + \varphi_f Y_{\text{fail}}$$

$$Y_{\text{fail}} = \begin{cases} 0 & P < 0 \\ \beta P & 0 \leq P \leq \overline{Y}/b \\ \overline{Y} & P \geq \overline{Y}/b \end{cases}$$

Fig. 7. Modified Wilkins' ceramic model: strength after fracture.

The penetration-time results for an impact of 820 m/s (the ballistic limit determined by Wilkins) are shown in Fig. 8, using values of $\beta = 1.4$ with no cap to limit the flow stress. This actually looks quite good. Since approximately 40% of the core is eroded in the simulation, the residual kinetic energy is ~0.50% of the initial kinetic energy. Although this would appear to be very good agreement, the model could not reproduce a number of experimental observations: it did not provide good estimates of residual velocities for overmatched impact conditions, it did not provide good estimates of residual projectile length, and it did not adequately predict dwell.

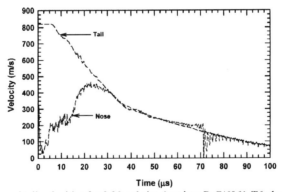

Fig. 8. Nose and tail velocities for 0.30-cal simulant into B₄C/6061-T6 aluminum target.

Hundreds of computer simulations were conducted, varying the parameters in the modified Wilkins' model. Particularly useful data were the residual length of the cores of the APM2 bullet, Fig. 9, since the core is only eroded by the ceramic (the core penetrates the aluminum substrate as a rigid body). Other very useful data were tests conducted for SwRI by the Army Research Laboratory using their 1-MeV flash X-ray system [12]. The targets, 7.62-mm B₄C tiles glued to 6.6-mm 6061-T6 aluminum substrates, were fabricated at SwRI. Three of the radiographs are shown in Fig. 10 at different times after impact. Dwell persists for approximately 20 µs; the bullet is just beginning to penetrate into the ceramic in Fig. 10(b). In order to match the various experimental data, the parameter $f_1$ in the Wilkins' model had to be decreased from 0.5 to 0.025. This parameter is associated with how fast the damage propagates through a computational cell. The original interpretation was that a crack propagated at some fraction of the sound speed. But to match the experimental data, $f_1$ had to be reduced by a factor of 20. This low value for $f_1$ was interpreted as the time it takes to comminute the ceramic. Hence, the somewhat now famous expression—"Cracks don't matter!"—which was uttered by the author at a ceramics workshop held at the Institute for Advanced Technology circa 1998. A comparison of the damage at 5 µs after impact for the two values of $f_1$ is shown in Fig. 11.

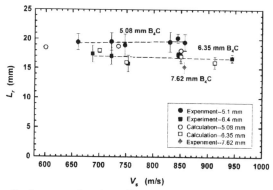

Fig. 9. Length of core as a function of impact velocity and ceramic tile thickness.

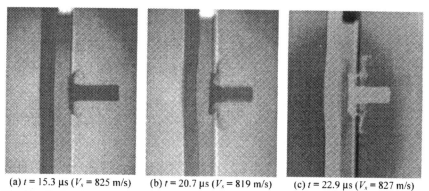

(a) $t$ = 15.3 µs ($V_s$ = 825 m/s)  (b) $t$ = 20.7 µs ($V_s$ = 819 m/s)  (c) $t$ = 22.9 µs ($V_s$ = 827 m/s)

Fig. 10. Flash radiographs of APM2 impacting a 7.62-mm $B_4C$/6.6-mm 6061-T6 Al target at approximately 825 m/s [12].

A comparison of a CTH simulation using the modified Wilkins' model to the data from the ARL experiments is shown in Fig. 12. The solid lines are the results of the simulation, and the solid points are the positions of various surfaces or interfaces measured from the flash radiographs. Agreement is very good.

So what had we learned about ceramics modeling, at least for relatively thin tiles backed by a ductile substrate material? Firstly, it is necessary to have a description of the strength of the failed material if the experimental data are to be reasonably reproduce; and secondly, it is the comminution of the ceramic—not crack propagation—that is important for local penetration.

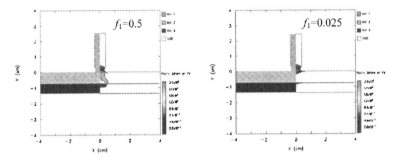

Fig. 11. Comparison of damage profiles at 5 µs for different values of $f_1$.

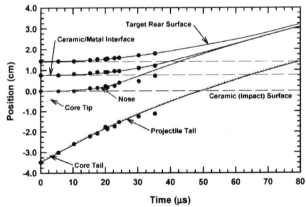

Fig. 12. Position-time for APM2 bullet impacting B₄C/6061-T6 target [12].

OTHER COMPUTATIONAL CERAMICS MODELS

The Wilkins (1968) and the modified Wilkins' ceramic models (1997-'98) are a tensile failure model, and are applicable to thin ceramic tiles. Rajendran [13] provides an excellent overview of other computational ceramic models. Only two will be discussed here—the Rajendran-Grove model and the Johnson-Holmquist model—but with more emphasis on the latter model. In the 1989-1990 timeframe, two ceramic models were developed, one by Rajendran, and the other by Johnson and Holmquist. The Rajendran-Grove model, based on micro-mechanics, degrades the elastic constants as damage accumulates in the form of microcracks [14-17]. Microcrack damage is measured in terms of a dimensionless microcrack density; the evolution law for growth/extension of the microcracks is derived from fracture mechanics and is based on a relationship for a single crack propagating under dynamic loading conditions. The ceramic "softens" (the elastic constants are degraded) as a function of the dimensionless microcrack density until some critical value of damage is reached, whereupon, the response is characterized by completely failed material. Prior to pulverization, the model also allows for pore collapse—to account for initial porosity in the intact ceramic—during compressive loading due to local microplastic flow of the matrix material surrounding the pores. Rajendran initially was interested in reproducing wave profiles from shock-wave (uniaxial strain flyer-plate) experiments,** but later began to apply the model to penetration, e.g., Ref. [14,17].

There was emphasis in using ceramics for heavy armor under the auspices of the DARPA heavy armor research initiative in the mid to late 1980's. Johnson and Holmquist developed a phenomenological computational ceramics model [18] under this effort, including how to determine constitutive constants [19]. The Johnson-Holmquist (J-H) model describes the "yield surface" for inelastic strain. The strengths of the intact and failed materials are functions of the confining pressure. The transition from intact to failed material is dependent on accumulated inelastic strain, which is also a function of pressure. Today, the J-H model is probably the most widely used computational ceramics model, although the form of, and the constants used in, the J-H model have evolved (and continue to evolve) [20-25]. The most recent model is shown in Fig. 13. The left-hand figure shows the equivalent stress for the intact and failed material as a function of confining pressure at two strain rates. The accumulation of damage, as measured by plastic strain, is shown in the top right figure. Bulking is denoted by the bottom right figure.

An analytical ceramics model was developed in 1996 for application to thin tiles [26]. The model assumes that the material being penetrated has failed and can be described by a Drucker-Prager yield surface. Two regions are envisioned: an inner region described by a pressure-dependent yield surface, which is surrounded by intact material (the second region). This required solution of an integral equation to define the boundary between the two regions. Further development of the model added a cap, which then required an interior boundary solution [27]. The pressure-dependent region corresponds to comminuted pieces sliding over each other; the cap corresponds to material-deforming plastic flow. The model was applied to long-rod penetration into semi-infinite ceramic targets [28-30]. Although this model is an analytical, in contrast to a computational, constitutive model, it is described here because it highlights one of the most pressing issues in computational ceramic modeling, which is the strength of the failed ceramic, the topic of the next section.

THE FAILED SURFACE

The constitutive form for the response of the failed material—a Drucker-Prager model—is common to all the models described above. Clearly, then, the response of the failed material is an extremely important aspect of any ceramic model. Walker took the constitutive constants for $B_4C$, as determined for the modified Wilkins' model from 100's of CTH simulations, and calculated the penetration of tungsten long rods into semi-infinite $B_4C$ targets as a function of impact velocity. The constitutive constants used were the slope $\beta = 1.7$ and the cap $\overline{Y} = 4.0$ GPa. The model results are compared to experimental data by Orphal, et al. [31] in Fig. 14.

---

** It was quickly learned that matching shock-wave profiles, although a necessary condition for the models, is not sufficient. Shock-wave profiles are insensitive to variations in some of the constitutive parameters; whereas, these parameters can be very important in penetration problems.

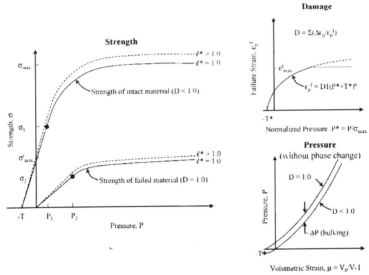

Fig. 13. Johnson-Holmquist-Beissel (JHB) model [25] (courtesy of T. Holmquist).

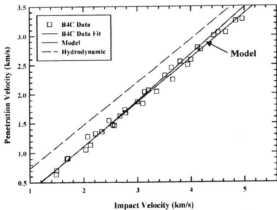

Fig. 14. Tungsten into $B_4C$: comparison of the Walker-Anderson ceramics model to experimental data from [28].

Agreement is quite good, except for the highest impact velocities. Earlier, during the ultra-lightweight armor program (but independent of the Wilkins' model calibration effort), Johnson and Holmquist developed constitutive constants for $B_4C$ for a slightly revised version of the J-H model (JH-2), which uses analytic equations to describe the yield surfaces. The equivalent stress vs. pressure for the intact and failed surfaces are shown in Fig. 15. The Drucker-Prager model parameters used by Walker are also plotted in the figure. The agreement between the J-H model parameters and those used by Walker are in very good agreement, although the actual position of the cap is in the "eye of the beholder." It was very satisfying that the constitutive parameters inferred from CTH simulations agreed so well with the experimental data from Wilkins, and Mayer and Faber, for failed material.[***] To place these results in historical perspective, the modified Wilkins' parameters were determined in 1997-'98, the JH-2 model parameters were developed in 1998-'99, and the Walker results, 2002.

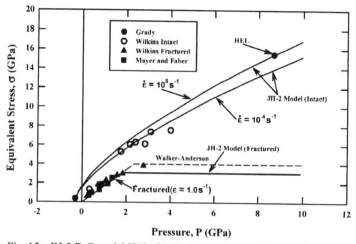

Fig. 15. JH-2 B₄C model [21] with Walker-Anderson failure surface (JH-2 model courtesy of T. Holmquist).

Walker determined the Drucker-Prager constitutive parameters for SiC by conducting parametric studies and comparing the results to experimental data of tungsten long rods impacting semi-infinite SiC targets [32]. He determined that the constitutive should be: $\beta = 2.5$; $\overline{Y} = 3.7$ GPa. Results are shown in Fig. 16. Further, he demonstrated that the slope and the cap were important, as shown in Fig. 17 [29]. The slope affects the penetration response (as compared to solely a constant flow stress) at the lower impact velocities, while the cap limits the effect of large penetration pressures at the higher impact velocities.

Holmquist and Johnson determined parameters for SiC [22], using the JH-1 model; the results are shown in Fig. 18. As there were no data for failed SiC material, they inferred the strength from some ballistic experiments, to be described in a few paragraphs. The J-H parameters for the failed material are: $\beta = 0.40$, $\overline{Y} = 1.3$ GPa. The description of the failed surface for the J-H model and the Walker-Anderson model are drastically different, as shown in Fig. 18. The time frame for the J-H results for SiC is 2001-'02.

---

[***] Because of the analytic form of the JH-2 model, the slope is not singled valued, but $\beta \approx 1.9$, compared to 1.7 for the Walker-Anderson (and modified Wilkins') model. The cap is 3.09 GPa in the JH-2 model.

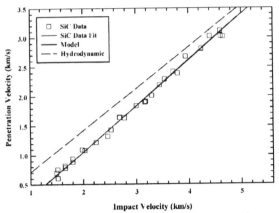

Fig. 16. Tungsten into SiC: comparison of the Walker-Anderson ceramics model to experimental data from [30].

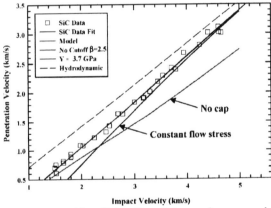

Fig. 17. Tungsten long rod into SiC: effect of slope and cap on model results [29].

But it gets worse! Some major enhancements were made to the J-H model. First, an algorithm for converting failed elements to generalized particles was developed, which suppressed numerical "noise" and maintained pressure at contact boundaries [33]. Additionally, they developed a capability to put a ceramic target under a state of prestress [24]. These capabilities had a dramatic influence on the magnitude of the cap for the failed surface.[****]

With the new capabilities, Holmquist and Johnson [25] re-examined experiments conducted by Lundberg, et al. [37]. These experiments, done in the reverse ballistics mode, involve launching highly confined ceramic targets at a very long (80-mm length, 2-mm diameter)

---

[****] Another capability added is the ability to treat solid-solid phase transitions [34-36]

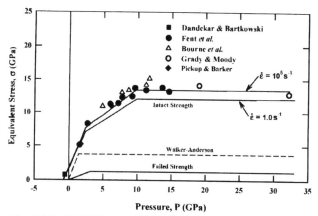

Fig. 18. Original J-H SiC model [22] with Walker-Anderson failure surface
(J-H model courtesy of T. Holmquist).

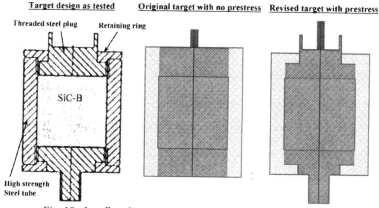

Fig. 19. Lundberg's target geometry (courtesy of T. Holmquist).

tungsten-alloy penetrator. The Lundberg target is shown on the left-hand side of Fig. 19; the SiC-B designation denotes a specific type of silicon carbide manufactured by Cercom. In the original work [22], Holmquist modeled the target as shown in the center figure of Fig. 19. In the more recent work [25], the geometric description of the target has higher fidelity, and included a prestress similar to what Lundberg used for his targets. The experimental data, along with computational results, are shown in Fig. 20. (The solid lines connecting the actual data points are simply straight lines that connect the data points to show the penetration-time trends.)

The procedure for estimating the response of the failed material remained the same. The test at an impact velocity of 1645 m/s (Fig. 20), where dwell persists for approximately 20 μs, is used to determine the initiation of damage (the accumulated inelastic strain). This sets the failure

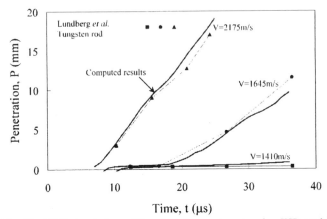

Fig. 20. EPIC simulations of Lundberg's experiments using JHB model
(courtesy of T. Holmquist).

strain. The material is intact and has the strength of the intact material while damage (plastic strain) is accumulating. Upon reaching the failure strain, the material then is assumed to fail suddenly, and the response switches from the intact surface to the failed surface. After failure, the cap of the failed surface is adjusted so that the numerically calculated penetration-time response of the 1645-m/s impact matches that of the experiment. This gives a maximum value for the failed material of 0.2 GPa, which is shown in Fig. 21. The calculated curves at impact velocities of 2175 m/s and 1410 m/s in Fig. 20 use the constitutive parameters determined from the 1645-m/s impact. It is seen that the calculations predict total interface defeat at 1410 m/s, and that they show the inflection points in the penetration-time response for the 2175-m/s impact (caused by having different penetration velocities resulting from time delays in ceramic failure). The equivalent stress vs. pressure for the new model is compared to that of for the older model in Fig. 21. (The original work used the JH-1 model, which uses straight-line segments for the yield surfaces. The new work used the JHB model, which has an analytic form for the yield surfaces.)

The Drucker-Prager model, using the constants derived by Walker for SiC, has been used in CTH to reproduce the experimental data of Orphal, et al. [38]. These data are also reproduced using the JHB model parameters for SiC [39]. Clearly, the discrepancies between the values by Walker for SiC are dramatically at odds with those used by the J-H model. One interpretation is that the Walker-Anderson model constants represent an "averaging" of the intact and failed material response. This would seem to make some sense, but this interpretation does not allow for the relatively good agreement between the constitutive constants for failed B₄C material.

It is highly desired that the constitutive constants be determined by independent laboratory experiments. SwRI has developed experimental techniques to determine response of confined SiC powder and *in-situ* damaged (comminuted) SiC [40]. The results are shown in Fig. 22, and compared to the J-H and Walker-Anderson model constants. Although an estimate of the cap has been made, there is considerable uncertainty in its value since the platens fail at the higher stress levels. There remain questions concerning the experimental procedures and the

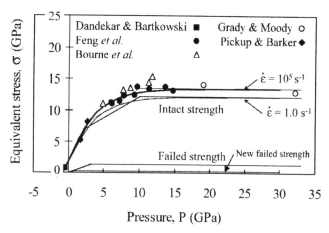

Fig. 21. Comparison of old and new SiC models with new failure surface (courtesy of T. Holmquist).

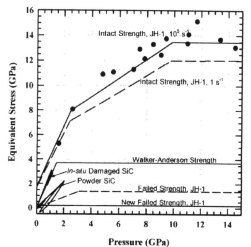

Fig. 22. Strength of comminuted and powder SiC tests compared to J-H and Walker-Anderson model parameters [40].

interpretation of the results, and on-going work is examining assumptions and interpretations. Additional research is also being conducted to refine and modify the experimental procedures, and new analysis tools are in development for the interpretation of the laboratory data. Nevertheless, the laboratory data are in approximate agreement with the Walker values. In fact, the strength of compacted SiC powder greatly exceeds that required by the JHB model to reproduce Lundberg's experiments. Thus, notwithstanding the demonstrated success of the J-H model, questions remain.

SUMMARY

An objective of computational constitutive models is to match as many diverse and types of experiments as possible. The geometry and/or impact conditions will change for these different experiments, but the same set of constitutive model parameters should be used. This has been demonstrated quite successfully with the Johnson-Holmquist ceramics model. However, another objective is that model parameters be determined from independent laboratory characterization experiments, i.e., non-ballistic impact experiments. To date, in order to match some experimental data, some model parameters—in particular, the initiation criterion for failure and the strength of the failed material—have been determined from ballistic experiments. The J-H parameters for the strength of failed material differ considerably than those inferred by Walker (also determined by matching experimental data) and direct laboratory characterization experiments of compacted ceramic powder and *in-situ* comminuted ceramic. To complicate things even further, it is often difficult to separate numerical hydrodynamics from the constitutive model, as already mentioned for the J-H model in the conversion of elements to particles [33] (also see [41]). Work is on-going to develop revised constitutive parameters for $B_4C$ [39], and which may have to also include the effects of phase transformations, e.g., [42]. Considerable discussion has ensued about the nature of the failed surface, and it is hoped that further research will sort these differences out.

REFERENCES

[1]M. Wilkins, C. Honodel, and D. Sawle, "An approach to the study of light armor," UCRL-50284, Lawrence Livermore National Laboratory, Livermore, CA (1967).

[2]M. L. Wilkins, "Second Progress Report of Light Armor Program," UCRL-50349, Rev. 1, Lawrence Livermore National Laboratory, Livermore, CA (1976).

[3]M. L. Wilkins, "Third Progress Report of Light Armor Program," UCRL-50460, Lawrence Livermore National Laboratory, Livermore, CA (1968).

[4]M. L. Wilkins, C. F. Cline, and C. A. Honodel, "Fourth Progress Report of Light Armor Program," UCRL-50694, Lawrence Livermore National Laboratory, Livermore, CA (1969).

[5]M. L. Wilkins, R. L. Landingham, and C. A. Honodel, "Fifth Progress Report of Light Armor Program," UCRL-50980, Lawrence Livermore National Laboratory, Livermore, CA (1971).

[6]M. L. Wilkins, "Calculations of Elastic-Plastic Flow," in *Methods of Computational Physics*, Vol. 3 (B. Adler, S. Fernback, and M. Rotenberg, Eds.), Academic Press (1964).

[7]M. L. Wilkins, "Calculation of Elastic-Plastic Flow," UCRL-7322, Rev. 1, Lawrence Livermore National Laboratory, Livermore, CA (1969).

[8]J. E. Reaugh, Lawrence Livermore National Laboratory, personal communication.

[9]J. D. Walker and C. E. Anderson, Jr., "The Wilkins' Computational Ceramic Model for CTH," SwRI Report 4391/002, Southwest Research Institute, San Antonio, TX (1991).

[10]J. M. McGlaun, S. L. Thompson, and M. G. Elrick, "CTH: A three-dimensional shock wave physics code," *Int. J. Impact Engng.*, **10**, 351-360 (1990).

[11]C. E. Anderson, Jr. and J. D. Walker, "Ceramic Dwell and Defeat of the 0.30-Cal AP Projectile," *15th U.S. Army Symp. on Solid Mech.*, Myrtle Beach, SC, April 12-14 (1999).

[12]C. E. Anderson, Jr., M. S. Burkins, J. D. Walker, and W. A. Gooch, "Time-Resolved Penetration of $B_4C$ Tiles by the APM2 Bullet," *Computer Modeling in Engng. & Science*, **8**(2), 91-104 (2005).

[13]A. M. Rajendran, "Historical Perspective on Ceramic Materials Damage Models," *Ceramic Transactions, Ceramic Armor Materials by Design*, (J. McCauley, *et al.*, Eds.), **134**, pp. 281-297, The American Ceramic Society, Westerville, OH (2002).

[14]A. M. Rajendran, "Modeling the Impact Behavior of AD85 Ceramic Under Multiaxial Loading," *Int. J. Impact Engng.*, **15**(6), 749-768 (1994).

[15]A. M. Rajendran and D. J. Grove, "Determination of Rajendran-Grove Ceramic Constitutive Model Constants," in *Shock Compression of Condensed Matter—1995* (S. C. Schmidt, Ed), pp. 539-542, AIP Press, NY (1996).

[16]A. M. Rajendran and D. J. Grove, "Modeling the Shock Response of Silicon Carbide, Boron Carbide, and Titanium Carbide," *Int. J. Impact Engng.*, **18**(6), 611-631 (1996).

[17]D. J. Grove and A. M. Rajendran, "Overview of the Rajendran-Grove Ceramic Failure Model", *Ceramic Transactions, Ceramic Armor Materials by Design*, (J. McCauley, *et al.*, Eds.), **134**, pp. 371-382, The American Ceramic Society, Westerville, OH (2002).

[18]G. R. Johnson and T. J. Holmquist, "A Computational Constitutive Model for Brittle Materials Subjected to Large Strains, High Strain Rates, and High Pressures," in *Shock Waves and High-Strain Rate Phenomena in Materials* (M. A. Meyers, L. E. Murr, and K. P. Staudhammer, Eds.), pp. 1070-1077, Marcel Decker, NY (1991).

[19]G. R. Johnson, T. J. Holmquist, J. Lankford, C. E. Anderson, and J. Walker, "A Computational Constitutive Model and Test Data for Ceramics Subjected to Large Strains, High Strain Rates, and High Pressures," Report DE-AC04-87AL-42550/1, Honeywell, Inc., Brooklyn Park, MN, August (1990).

[20]G. R. Johnson and T. J. Holmquist, "An Improved Computational Constitutive Model for Brittle Materials," in *High-Pressure Science and Technology—1993*, (S. C. Schmidt, J. W. Shaner, G. A. Samara, and M. Ross, Eds.), 981-984, AIP Conf. Proc. 309, AIP Press, NY (1994).

[21]G. R. Johnson and T. J. Holmquist, "Response of Boron Carbide Subjected to Large Strains, High Strain Rates, and High Pressures," *J. Appl. Phys.*, **85**(12), 8060-8073 (1999).

[22]T. J. Holmquist and G. R. Johnson, "Response of Silicon Carbide to High Velocity Impact," *J. Appl. Phys.*, **91**(9), 5858-5866 (2002).

[23]T. J. Holmquist and G. R. Johnson, "Ceramic Dwell and Interface Defeat," *Ceramic Transactions, Ceramic Armor Materials by Design*, (J. McCauley, *et al.*, Eds.), **134**, pp. 485-498, The American Ceramic Society, Westerville, OH (2002).

[24]T. J. Holmquist and G. R. Johnson, "Modeling Prestressed Ceramic and its Effects on Ballistic Performance," *Int. J. Impact Engng.*, **31**, 113-127 (2005).

[25]T. J. Holmquist and G. R. Johnson, "Characterization and Evaluation of Silicon Carbide to High Velocity Impact," *J. Appl. Phys.*, **97**(09), 093502/1-12 (2005).

[26]J. D. Walker and C. E. Anderson, Jr., "An Analytic Model for Ceramic-Faced Light Armor," *Proc. 16th Int. Symp. on Ballistics*, Vol. 3, pp. 289-298, San Francisco, CA (1996).

[27]J. D. Walker and C. E. Anderson, Jr., "An analytic penetration model for a Drucker-Prager yield surface with cutoff," *Shock Compression of Condensed Matter—1997*, S. C. Schmidt, Eds., 897-900 (1998).

[28]J. D. Walker, "Analytic Model for Penetration of Thick Ceramic Targets," *Ceramic Transactions, Ceramic Armor Materials by Design*, (J. McCauley, *et al.*, Eds.), **134**, pp. 337-348, The American Ceramic Society, Westerville, OH (2002).

[29]J. D. Walker, "Influence of Drucker-Prager Constitutive Parameters on Penetration of Thick Ceramic Targets," *Proc. 20th Int. Symp. on Ballistics*, **2**, 794-801, DES*tech* Publications, Lancaster, PA (2002).

[30]J. D. Walker, "Analytically Modeling Hypervelocity Penetration of Thick Ceramic Targets, *Int. J. Impact Engng.*, **29**, 747-755 (2003).

[31]D. L. Orphal, R. R. Franzen, A. C. Charters, T. L Menna, and A. J. Piekutowski, "Penetration of Confined Boron Carbide Targets by Tungsten Long Rods at Impact Velocities from 1.5 to 5.0 km/s, *Int. J. Impact Engng.*, **19**(1), 15-29 (1997).

[32]D. L. Orphal and R. R. Franzen, "Penetration of Confined Silicon Carbide Targets by Tungsten Long Rods at Impact Velocities from 1.5 to 4.6 km/s," *Int. J. Impact Engng.*, **19**(1), 1-13 (1997).

[33]G. R. Johnson, S. R. Beissel, and T. J. Holmquist, "Improved Computational Modeling of Ballistic Problems using Meshless Particles and Finite Elements," *Proc. 20th Int. Symp. on Ballistics*, **2**, 850-858, DES*tech* Publications, Lancaster, PA (2002).

[34]T. J. Holmquist, D. W. Templeton, and K. D. Bishnoi, "Constitutive Modeling of Aluminum Nitride for Large Strain, High-Strain Rate, and High-Pressure Applications," *Int. J. Impact Engng.*, **25**(3), 211-232 (2001).

[35]T. J. Holmquist, D. W. Templeton, and K. D. Bishnoi, "High Strain Rate Constitutive Modeling of Aluminum Nitride including a First-Order Phase Transition," *J. Phys., IV*, France **10**, Pr9/21-26 (2000).

[36]G. R. Johnson, T. J. Holmquist, and S. R. Beissel, "Response of Aluminum Nitride (including a Phase Change) to Large Strains, High Strain Rates, and High Pressures," *J. Appl. Phys.*, **94**(3), 1639-46 (2003).

[37]P. R. Lundberg, R. Renstrom, and B. Lundberg, "Impact of Metallic Projectiles on Ceramic Targets: Transition Between Interface Defeat and Penetration," *Int. J. Impact Engng.*, **24**, 259-275 (2000).

[38]D. L. Orphal, C. E. Anderson, Jr., D. W. Templeton, Th. Behner, V. Hohler, and S. Chocron, "Using Long-Rod Penetration to Detect the Effect of Failure Kinetics in Ceramics," *Proc. 21st Int. Symp. on Ballistics*, pp. 744-751, Adelaide, Australia, 19-13 April (2004).

[39]T. J. Holmquist, personal communication.

[40]S. Chocron, K. A. Dannemann, A. E. Nicholls, J. D. Walker, and C. E. Anderson, Jr., "A Constitutive Model for Damaged and Powder Silicon Carbide," *Proc. 29th Int. Conf. Advanced Ceramics & Composites, Ceramic Engng. & Sci. Proc.*, **26**(7), 35-42, American Ceramics Society (2005).

[41]C. E. Anderson, Jr., I. S. Chocron, and C. E. Weiss, "Analysis of Time-Resolved Penetration of Long Rods into Glass Targets—II," *30th Int. Conf. & Exp. on Advanced Ceramics and Composites*, The American Ceramics Society, Cocoa Beach, FL, 22-27 Jan. (2006).

[42]M. Chen, J. W. McCauley, and K. J. Hemker, "Shock-Induced Localized Amorphization in Boron Carbide," *Science*, **299**, 1563-1566 (2003).

# Silicon Carbide

# BIOMORPHIC SISIC-MATERIALS FOR LIGHTWEIGHT ARMOUR

Bernhard Heidenreich, Michaela Gahr
DLR – German Aerospace Center
Institute of Structures and Design
Pfaffenwaldring 38-40
D-70569 Stuttgart

Elmar Straßburger
Fraunhofer Gesellschaft e.V.
EMI – Ernst Mach Institute
Am Klingelberg 1
D-79588 Efringen-Kirchen

Dr.-Ing. Ekkehard Lutz
ETEC Gesellschaft für technische Keramik mbH
An der Burg Sülz 17
D-53797 Lohmar

## ABSTRACT

Lightweight ceramic armour systems offer high ballistic protection combined with a significantly lower areal weight than conventional armour systems on the basis of steel or aluminum. Biomorphic SiSiC-ceramics, based on wooden preforms and manufactured via the liquid silicon infiltration (LSI), are considered promising new armour materials due to their high potential for the manufacture of complex shaped structures at low costs.

The physical properties of different, biomorphic SiSiC-materials were determined. The ballistic performance was studied on sample plates 100 mm x 100 mm with armour piercing ammunition (7.62 x 51 mm AP). The results were compared to the ballistic performance of commercially available alumina tiles, namely ALOTEC 96SB and 99SB, which were tested under the same conditions. In this paper, the manufacturing process of biomorphic SiSiC-materials, the material properties and the results of the impact tests are presented.

## INTRODUCTION

Over the many years up to now, alumina is considered the major standard low cost material, which is being used as hard face in armour composite systems in body and vehicle protection. In most cases, the alumina layer is made of tiles, e. g. 50 x 50 mm² or 100 x 100 mm². Whereas the typical thickness of such tiles varies from 4 to 11 mm in body armour, it can reach values of up to 25 mm in vehicle armour, always depending on threat and soft-armour backing material.

There are various commercial alumina grades available on the market that exhibit alumina contents between 92 to 99.7 % with densities from 3.65 to 3.91 g/cm². The prices range from less than USD 10 to about USD 20 per kg for superior grades, also depending on the complexity of layout of such composites.

There are a variety of silicon and boron carbide ceramics (SiC and $B_4C$) available, which are considerably more expensive than alumina, but which provide an inherently lower density of 3.2 down to 2.5 g/cm³, respectively. Assuming that these "black" ceramics provide similar ballistic performance at the same thickness, their weight contribution is by about 16 to 35 % less than alumina.

For silicon carbide armour materials reaction-bonded SiC (RBSC or SiSiC; price from USD 35/kg) and sintered SSiC and LPSSC (prices from USD 80/kg) is available. Boron carbide comes as RB $B_4C$ and hot pressed $B_4C$ at costs of at least USD 150 per kg. Hence, there is a price to performance gap between the "white" alumina ceramics and the black carbide materials that still waits to be filled.

In the work presented, novel SiSiC materials were investigated, which have a high economic potential due to their cost effective manufacturing process. This so called LSI process (liquid silicon infiltration) was originally developed for carbon fibre reinforced ceramic matrix composites (CMC), so called C/C-SiC materials, used for thermal protection systems of spacecraft [1]. Modified low cost material variations with short fibre reinforcement could be transferred successfully to industrial applications, e.g. high performance brake disks [2].

For the manufacture of biomorphic SiSiC materials, based on low cost green bodies made of wood, the LSI process has been applied successfully (Fig. 1) [3].

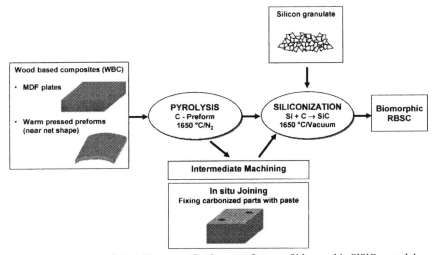

Fig.1: Schematic view of the LSI-process for the manufacture of biomorphic SiSiC materials.

For the green bodies, commercially available medium dense fibre boards (MDF), as well as wood based composites (WBC), specially developed by German Aerospace Center (DLR) are used. MDF panels are widely used in the furniture industry and are made by pressing fine fibres of needle wood with phenolic resin in a mass production process. For the WBC materials, powder of wood and carbon rich precursors, such as phenolic resins, are used but the composition, density, open porosity and microstructure of these green bodies are adapted to the LSI process, leading to different qualities of biomorphic SiSiC materials [4]. Flat panels as well as parts with complex geometries are manufactured at DLR via warm pressing in a batch process.

In the second step of the LSI process, the wood based green body is pyrolysed at maximum temperatures of 1650 °C in nitrogen atmosphere. In the last process step, the resulting carbon preform, characterized by a high, open porosity of 40 - 60 %, is siliconized at 1650 °C in vacuum. Thereby molten silicon is infiltrated by capillary action into the C-preform, reacting with carbon to form SiC. The resulting biomorphic SiSiC is a dense multiphase material consisting of SiC and Si with generally very small amounts of C. The phase contents, and therefore the properties, can be varied in a wide range by the use of different raw materials and process parameters. The SiC-content, for example, can be varied between 20 % and more than 85 % by volume. Due to a unique in-situ joining technology, reproducible shrinkage rates during pyrolysis, and the geometrical stability during the siliconization, even large and complex shaped parts can be manufactured via LSI in near net shape technique, and therefore waste and machining costs can be minimized (Fig. 2). The maximum dimensions of biomorphic SiSiC parts are mainly restricted by the available furnace room, and by the thickness of the components. At DLR, plates up to 335 mm x 335 mm with a thickness of up to 50 mm have been manufactured until now.

In this study, the ballistic performance of these relatively new, biomorphic SiSiC materials, was investigated and compared to commercially available alumina. In a previous work it was demonstrated how well these biomorphic SiSiC materials performed, when compared to different carbon fibre reinforced C/C-SiC ceramic composites [5].

Fig.2: Examples of components, manufactured via LSI in near net shape technique. Top left: Nose cap for X38 manufactured from C/C-SiC (ca. 760 mm x 660 mm x 190 mm, d = 6 mm); Top right: short fibre reinforced C/C-SiC brake disc for automobiles; Bottom left: in situ joined sample parts and crucible from biomorphic SiSiC ( 340 mm x 120 mm x 70 mm, d = 10 mm); Bottom right: ring elements (∅ 140 mm, d = 20 mm) from biomorphic SiSiC on the basis of machined C-preforms.

## MANUFACTURE OF CERAMIC TILES

In this work, nonoxide as well as oxide ceramic materials were investigated. The two different types of nonoxide materials were based on biomorphic SiSiC, manufactured via the LSI process at DLR. For the oxide materials, two types of alumina tiles, ALOTEC 96SB and ALOTEC 99SB, manufactured by dry pressing and sintering were used. These commercially available tiles are produced in high volumes by ETEC GmbH, Germany, and are successfully used for various lightweight armour applications like body armour or vehicle protection systems.

Biomorphic SiSiC ceramics were manufactured either by the use of commercially available medium dense fibre boards (MDF) as green bodies or from wood based composites (WBC), developed at DLR. In order to achieve a near net shape manufacturing of the sample plates, the MDF was cut to small sample plates (130 mm x 130 mm x 21 mm). To remove the moisture content, the plates were dried for 24 hours at 110 °C in air. For the WBC green bodies, milled wood powder (grain size < 30 μm) and powdery phenolic resin (grain size < 15 μm) were mixed in a dry process using an Eirich mixer (model RV 02 E). After filling in the press mass in a steel mould, the compound was uniaxially densified in a press and cured at a maximum temperature of 185 °C. The curing process of the phenolic resin is accompanied by the formation of water. To remove this absorptive integrated water content of about 5-8 mass-%, the resulting WBC-plates (335 mm x 335 mm x 15 mm) were dried for 24 hours at 110 °C in air after demoulding.

The WBC and MDF plates were pyrolysed in inert gas atmosphere at maximum temperatures of up to 1650 °C. Due to the transformation of both, the wood based fibres and powders as well as of the phenolic resin into carbon, the volume as well as the mass of the preforms was reduced significantly. The WBC plates showed slightly lower shrinkage rates of 61 % by volume and 64 % by mass than the MDF plates (64 volume-%, 70 mass-%). Due to linear shrinkages of 23 % (MDF) and 25 % (WBC) in the in plane direction and 40 % (MDF) / 30% (WBC) in the thickness direction, the geometries of the plates were reduced to 100 mm x 100 mm x 12.7 mm and 251 mm x 251 mm x 11 mm respectively. The resulting C-preforms showed relatively high open porosities of 58.8 % (MDF) and 43.6 % (WBC) and low densities of 0.6 g/cm³ (MDF) and 0.82 g/cm³ (WBC). Before the siliconization, the WBC based carbon preforms were cut to sample plates (100 mm x 100 mm) by diamond saw cutting.

All the carbon preforms were siliconized in vacuum, at maximum temperatures of about 1650 °C. Thereby silicon in form of granules was melted and infiltrated in the porous C preforms only by capillary forces. The geometry of the plates was not influenced by the siliconization. After cooling down, the biomorphic SiSiC tiles were ground to a thickness of 8 mm with a diamond coated tool.

Alumina armour tiles are generally made by dry pressing and sintering. Thereby, submicron alumina powder with low contents of binders and sintering aids is axially pressed in metal moulds to flat plates with maximum dimensions of around 300 mm x 360 mm (e.g. green size of so-called monoliths for body armour). After that, the green preforms can be cut to suit any flat layout. All standard and cut up tiles are then sintered at temperatures exceeding 1600 °C in air. Due to reproducible linear shrinkage rates of slightly less than 20 % during sintering, a net shape manufacturing is possible, and therefore, no machining of the final ceramic parts is necessary.

Two different alumina materials with alumina contents of 96 % (ALOTEC 96 SB) and 99 % (ALOTEC 99 SB) were investigated. ALOTEC 99 SB exhibits a maximum sinterable alumina content of 99.7 %, and consequently, only about 0.3 % of sintering aids. This leads to a material with slightly higher density and mechanical properties, compared to ALOTEC 96 (Table 1). Sintering leads to a closed porosity and a fine-grained homogeneous microstructure (Fig. 3).

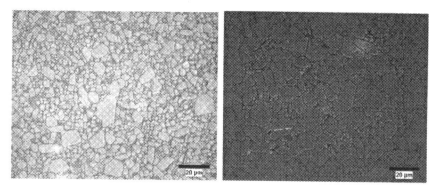

Fig. 3: SEM image of alumina armour materials ALOTEC 96 SB (left) and ALOTEC 99 SB (right) with a characteristic fine grained, homogeneous bimodal microstructure.

To compare the ballistic performance of alumina and SiSiC tiles, the alumina tiles were ground to a thickness of 6.3 mm (ALOTEC 96 SB) and 6.2 mm (ALOTEC 99 SB) with diamond coated tools, leading to the same areal weight of the target samples ($\approx$ 36 kg/m²) as was achieved with the targets, based on the SiSiC tiles with a thickness of 8 mm.

SAMPLE PREPARATION

For the ballistic tests, 22 targets were built, consisting of ceramic tiles of lateral dimensions 100 mm x 100 mm, glued with polyurethane adhesive (Sikaflex) onto Twaron T 750 aramid laminates. The T 750 laminates consisted of 20 fabric layers in all tests. Except the target with 8.2 mm ALOTEC 96 SB, all targets exhibited approximately the same total areal weight of $\approx$ 36 kg/m² (Table I).

Table I: Overview of tested targets with different ceramic tiles and aramid backing.

| Target | | Biomorphic SiSiC | | | Al₂O₃ | |
|---|---|---|---|---|---|---|
| | | | | | ALOTEC | |
| | | MDF | WBC | 96 SB | 96 SB | 99 SB |
| Tile thickness | [mm] | 8 | 8 | 8.2 | 6.3 | 6.2 |
| Number of aramid layers in backing | | 20 | 20 | 20 | 20 | 20 |
| Areal weight | [kg/m²] | 34 | 36.4 | 42.7 | 35.5 | 34.6 |
| Number of targets | | 4 | 4 | 3 | 5 | 6 |

TESTING

Material Properties of biomorphic SiSiC materials

The Microstructure was studied using scanning electron microscopy (SEM) for polished sections. Density and open porosity of all tiles was determined using the water immersion method based on the Archimedes law (DIN EN 993-1). Young's modulus was tested according to DIN EN 843-2, three-point flexural strength in accordance to ASTM C113. Rockwell hardness was determined according to DIN EN 10 109. Fracture toughness (critical stress intensity factor) $K_{Ic}$ was determined using SENVB samples according to ASTM STP 1409.

Ballistic tests

The ballistic performance of the different ceramic materials was tested with 7.62 mm x 51 AP FN armour piercing ammunition with a total projectile mass of 9.5 g and a steel core mass of 3.7 g. The impact velocity ($v_P$) was varied between 600 and 850 m/s in order to determine the ballistic limit velocities $v_{BL}$. The samples were clamped in a steel frame via the backing plate and each sample was tested with a single hit. The ballistic limit velocities were determined using the Lambert -Jonas approach: $v_R = [K(v_P^2 - v_{BL}^2)]^{1/2}$, with $v_R$ = residual velocity after perforation [6]. In this approach only the balance of the kinetic energy is considered. A curve of the Lambert-Jonas type delivers a good approximation of the experimental data. The parameter K and the ballistic limit velocities $v_{BL}$ were determined by a least squares fit of the experimental data.

RESULTS

The biomorphic SiSiC materials based on WBC showed a significantly higher ceramic content ($\varphi_{SiC}$ = 85 vol.-%) and a more homogeneous microstructure (Figure 4) compared to the MDF-based material ($\varphi_{SiC}$ = 61 vol.-%), resulting in a higher density and significantly higher mechanical properties (Table II).

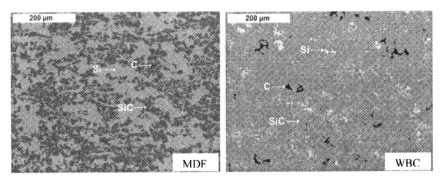

Fig. 4: SEM image (200 x, polished, cross section) of biomorphic SiSiC derived from MDF (left) and WBC (right) consisting of SiC (dark gray), C (black) and Si (light gray).

Table II: Mechanical properties of tested alumina and biomorphic SiSiC materials compared to conventional reaction bonded SiC.

| Material Properties | Unit | Biomorphic SiSiC | | Conventional RBSC [7] | Al$_2$O$_3$ ALOTEC [8] | |
|---|---|---|---|---|---|---|
| | | MDF | WBC | | 96 SB | 99 SB |
| Density | [g/cm³] | 2.8 | 3.1 | 3.0 – 3.17 | 3.8 | 3.93 |
| Open Porosity | [%] | 1.4 | 0.1 | - | 0 | 0 |
| SiC / Si content | [Vol.- %] | 61 / 35 | 85 / 13 | - | - | - |
| Al$_2$O$_3$ content | [Vol.- %] | - | - | - | 96.0 | 99.7 |
| Young's Modulus | [GPa] | 276 | 386 | 300 - 400 | ≥310 | ≥365 |
| Flexural Strength | [MPa] | 232 | 400 | 190 - 250 | ≥250 | ≥280 |
| K Ic | [MPam$^{0.5}$] | 2.6 | 2.8 | - | - | - |
| Hardness HRA | [HRA] | 77 | 83 | 95 | - | - |
| Vickers HV10 | [GPa] | - | - | 23 | ≥12 | ≥15 |
| Vickers HV1 | [GPa] | 20.1 | - | 23-24 | - | - |

The ALOTEC 96 tiles with a thickness of 8.2 mm and an areal weight of 42.7 kg/m² stopped the projectiles at maximum impact velocities of 840 m/s in all three tests (Fig.5). The alumina tiles with reduced thickness (6.2 / 6.3 mm) and areal weights of about 36 kg / m² could not stop the projectiles in most of the tests. The resulting ballistic limit velocity for ALOTEC 99 (735 m/s) was 5 % higher than that of ALOTEC 96 (700 m/s) despite a pronounced deviation. This is within the range of 5 to 10 % of previous observations during internal testing under various conditions.

The biomorphic SiSiC derived from MDF could stop the projectiles with impact velocities of up to 775 m/s. In one single test with an impact velocity of 834 m/s, the projectile penetrated the armour system with a residual velocity of $v_R$ = 465 m/s. Therefore, the ballistic limit velocity was calculated to 775 m/s for the MDF based biomorphic SiSiC.

Due to the higher SiC content, the WBC based biomorphic SiSiC showed a slightly higher ballistic performance compared to the MDF based material. With a tile thickness of 8 mm, three out of the four test targets stopped the projectiles, which had an average impact velocity of 832 m/s. One target was penetrated by the projectile with $v_R$ = 335 m/s. The ballistic limit velocity was calculated to 800 m/s. However, only four samples of this type of SiSiC were available and the $v_R$-results fell into the so-called zone of mixed results, where perforation and projectile defeat can occur. Therefore, the ballistic limit velocity could only be estimated, assuming a $v_R$-$v_P$-curve shape similar to the other targets.

In every case where the projectile was stopped, the hardened steel core was broken, but the ceramic tiles were strongly fragmented (Fig. 6). Due to the large deformation of the aramid laminates most of the ceramic fragments were separated from the backing.

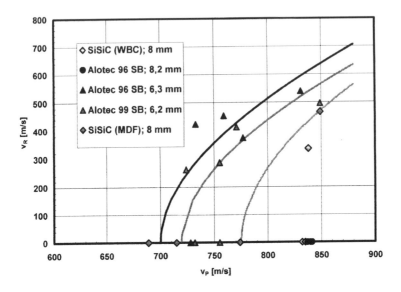

Fig. 5: Residual velocity $v_R$ versus impact velocity $v_P$ for the different sample targets, based on ceramic tiles with different thicknesses and on aramid backing.

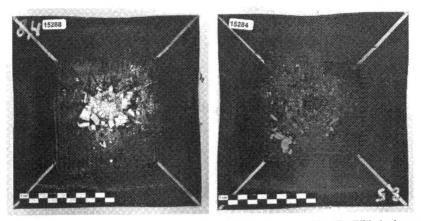

Fig. 6: Targets with aramid backing after impact with 7.62 mm x 51 AP. With both target samples the projectile was stopped and the hardened steel core was destroyed. Left: alumina tile (ALOTEC 96; d = 8.2 mm; $v_P$ = 841 m/s). Right: biomorphic SiSiC (WBC; d = 8 mm; $v_P$ = 832 m/s).

## DISCUSSION

The targets based on ALOTEC 96 SB tiles with a tile thickness of 8.2 mm showed the highest ballistic limit velocity ($v_{BL}$ = 840 m/s) of all tested targets (Fig. 7). The ballistic performance of this alumina tiles was directly influenced by the tile thickness. Reducing the tile thickness by 23.2 % to 6.3 mm lead to a decrease in the target areal weight of 11.8 % but also to a significant reduction of the ballistic limit velocity by 16.7 % ($v_{BL}$ = 700 m/s). With the higher quality ALOTEC 99 SB, the ballistic performance could not be improved at this low tile thickness.

With a tile thickness of about 8 mm, the ballistic limit velocity of the targets with biomorpic SiSiC based on WBC was on a comparable level to the commercially used ALOTEC 96 SB. However, due to the lower density of the SiSiC material, the resulting areal weight was 15 % lower and therefore these new SiSiC materials showed a higher ballistic performance for lightweight armour systems.

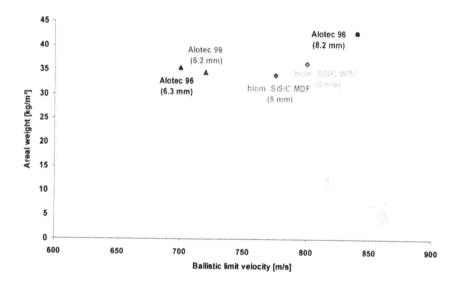

Fig. 7: Ballistic limit velocity in dependence of the areal weight of tested ceramic armour systems with aramid backing.

Compared to former results of internal investigations, this new biomorphic SiSiC materials showed a better ballistic performance than conventional reaction bonded, silicon infiltrated SiC materials and seems to be comparable even to liquid phase sintered SiC materials (LPSSC), commonly used for lightweight armour systems. This LPSSC is manufactured on the basis of fine ceramic powders of SiC and alumina at sintering temperatures above 2000 °C and pressures of 20 – 30 MPa.

Compared to the LPSSC, the biomorphic SiSiC materials has great advantages, regarding the manufacturing costs and the feasibility of complex structures, and therefore offers a high potential for the use in future lightweight armour systems. The manufacturing costs of biomorphic SiSiC materials are relatively low, because of the use of low cost raw materials such as wood powders, phenolic resin and silicon granulate from mass production. Using MDF preforms, the cost of the raw materials are determined by the silicon. For the manufacture of 1 kg of biomorphic SiSiC, the raw material costs are resulting from about 0.5 USD for the MDF preform and about 55 USD for the silicon, based on the currently used high purity Si with a price of 69 USD/kg. Further investigations will show whether the material costs can even be decreased by the use of cheaper Si-qualities which are offered today for 32 – 46 USD/kg for small batches. Additional advantages of the biomorphic SiSiC are the relatively simple ceramization process, with no pressure needed, and the lower process temperatures as well as the possibilities to manufacture large and complex shaped parts in near net shape technique.

SUMMARY

The LSI method is a cost effective process for the manufacture of fibre reinforced CMC materials as well as for structural ceramics on the basis of SiC. Large parts with complex geometries and curved shapes can be realized in near net shape technique, due to reproducible shrinkage rates and a unique in-situ joining method.

Two different types of biomorphic SiSiC, derived from wood based preforms (MDF and WBC), as well as two commercially used alumina armour materials with different alumina contents (ALOTEC 96, ALOTEC 99) were investigated. The target samples, built up with an aramid backing, were tested with single shots against armour piercing ammunition (7.62 x 51 mm AP).

The biomorphic SiSiC material derived from wood based composites (WBC) showed a 40 % higher SiC content compared to the biomorphic SiSiC based on medium dense fibre boards (MDF), leading to higher mechanical properties and higher ballistic performance. WBC-based SiSiC could stop the projectiles at maximum velocities of 835 m/s. Due to a penetration of one target, and because of the limited test data, the ballistic limit velocity was calculated as 800 m/s.

The ALOTEC 96 alumina tiles with a thickness of 8.5 mm stopped the projectiles at maximum projectile velocities of 840 m/s. However, the total areal weight (42.7 kg/m²) was 15 % higher compared to the targets equipped with WBC based SiSiC (36.4 kg/m²). With comparable areal weight, obtained by reduced tile thicknesses, the ballistic performance of both alumina materials was even lower than that of MDF based biomorphic SiSiC. A significant influence of the different qualities of alumina tiles to the ballistic limit velocity could not be observed with this specific test configuration.

With the biomorphic SiSiC based on WBC, a similar level of ballistic limit velocity as with the commercially used ALOTEC 96 SB could be achieved. Due to the lower density of the SiSiC material, the ballistic performance seems to be comparable to commonly used SiC armour materials. In comparison to liquid phase sintered LPSSC or hot pressed HPSC, the biomorphic SiSiC offers great economical advantages due to the low-cost manufacturing process via LSI, especially for large and complex shaped parts. Therefore, these new materials have a high potential for the use in lightweight armour for aircraft and land vehicles as well as for body armour applications.

Further tests are planned to verify the ballistic properties of biomorphic SiSiC material variations. The future development will be focused on modified SiSiC materials as well as on large armour plates with improved multi-hit performance as well as on more sophisticated backing systems.

REFERENCES

[1]W. Krenkel, "Entwicklung eines kostengünstigen Verfahrens zur Herstellung von Bauteilen aus keramischen Verbundwerkstoffen", *Dissertation Universität Stuttgart, DLR-Forschungsbericht* 2000-04, Stuttgart, (2000)

[2]W. Krenkel, R. Renz, B. Heidenreich, "Lightweight and Wear Resistant CMC Brakes", *7th International Symposium Ceramic for Engines*, Goslar, (2000)

[3]M. Gahr, J. Schmidt, W. Krenkel, A. Hofenauer, O. Treusch, "SiC-Keramiken auf der Basis von Holzwerkstoffen", *Verbundwerkstoffe (Eds.: H. P. Degischer)*, p. 383, WILEY-VCH, Weinheim, Germany (2003)

[4]M. Gahr, J. Schmidt, A. Hofenauer, O. Treusch, "Dense SiSiC ceramics derived from different wood-based composites: Processing, Microstructure and Properties", *Proceedings of the 5th International Conference on High Temperature Ceramic Matrix Composites (HTCMC 5)*, p. 425, The American Ceramic Society (2004)

[5]B. Heidenreich, B. Lexow, "Schmelzinfiltrierte SiC-Keramiken für Leichtbaupanzerungen", *Keramische Zeitschrift* 55, 10, p. 794 (2003) ,

[6]J.P. Lambert, G.H. Jonas, Ballistic Research Laboratory, Report BRL-R-1852, 1976

[7]B. Heidenreich, M. Gahr, E. Medvedovski, "Biomorphic reaction bonded silicon carbide ceramics for armour applications", *107th Annual Meeting & Exposition*, Baltimore (2005)

[8]ETEC Gesellschaft für technische Keramik mbH: Company, Material Data Sheets and Brochures, Lohmar (2005)

# EVALUATION OF SiC ARMOR TILE USING ULTRASONIC TECHNIQUES

J. Scott Steckenrider
Illinois College
1101 W. College Ave.
Jacksonville, IL 62650

William A. Ellingson, Rachel Lipanovich, Jeffrey Wheeler, Chris Deemer
Argonne National Laboratory
9700 S. Cass Avenue
Argonne, IL 60439

## ABSTRACT

Obtaining reliable ballistic performance is important for full utilization of ceramic armor tile. In this work, various sets of armor tiles were studied using multiple ultrasonic approaches. One set of specimens consisted of three specially prepared SiC armor tile, 10 cm by 10 cm by 1.9 cm thick, and seeded with various defects representing those most likely to occur and cause ballistic failure. Water-coupled ultrasound studies were performed using both a conventional water-coupled system (in which time-of-flight measurements were used instead of conventional C-scan parameters) and a phased-array water-coupled system. It was demonstrated that while the conventional system, along with advanced digital image processing, allowed detection of most of these defects, the phased-array system achieved comparable if not superior results in a fraction of the time. This paper will present the test specimens that have been used so far in these studies, the details of the data acquisition parameters and the results obtained to date.

## INTRODUCTION

Poor ballistic performance of SiC armor tile for vehicles is a concern for reliable protection of military personnel. An effective non-destructive screening technique to sort "good" tile from "bad" tile, assuming such knowledge exists, would be one possible method to reduce ballistic performance failures[1]. One likely source of this differentiation would be the presence of flaws, such as inclusions of differing density or large-grain regions[2]. Conventional ultrasonic methods have been evaluated as a tool for identifying and locating these flaws (from which subsequent correlation to ballistic performance would be established) and have demonstrated promise in preliminary work[3,4]. However, these were not applied to "realistic" defects and required a significant investment of scan time. Phased array ultrasonic technology has been suggested as an alternative method that would improve not only sensitivity but also throughput[3,5], as it would allow fast scanning over a large region with high Signal-to-Noise Ratio (SNR) for defect depth discrimination and would likely be cost effective[6,7].

## APPROACH

To evaluate the detectability of these seeded representative defects by conventional ultrasonic methods each of the three tiles was scanned first using Argonne's water-coupled SONIX ultrasonic system. A conventional ultrasonic focused transducer from Olympus NDT (formerly R/D Tech and Panametrics NDT) that was 19mm in diameter, had a nominal 12.7 cm focal length in water and operated at 5 MHz was used for this analysis. All data were recorded using an A/D interface operating at a double sampled 400 MHz acquisition rate. For a more detailed description of the ultrasonic methods used, the reader is referred elsewhere[3]. The tiles were then inspected using the phased array system, a TomoScan Focus LT by Olympus NDT. The phased array probe consists of 128 transducer elements operating at 10MHz. The active area is 64 mm by 7 mm. Prior work has demonstrated that SNR improves with increasing frequency for the SiC materials under consideration up to a maximum around 20 MHz[3]. However, due to probe availability limitations, the initial phased array analysis was performed at 10 MHz. Both the conventional and phased array analyses were performed at a fixed focus roughly 3 mm from the back surface of the tile (where the defect seeding was to have been applied). Finally, to evaluate the performance of the phased array system when defect depth is initially unknown, a volumetric "depth scan" was performed using the phased array system in which the focus is scanned in all three dimensions.

TEST SAMPLES

In order to evaluate "realistic" defects, three tiles with intentional defects were produced by a separate company under contract to the US Army Research Laboratory ( US ARL), Weapons and Materials Research Directorate (WMRD), Survivability Branch and were produced as part of a larger program. The "designed" defects were to represent those that could occur in normal processing. The sizes of the defects introduced were within the range of what could be of interest to those involved with ballistic testing. The designed defects in the tile included : a) inclusions of a lower-density material (seeded in tile "P"), b) inclusions of a higher-density material (seeded in tile "V") and c) larger sized grains (seeded in tile "A"). In each tile, the defects were seeded in a layer 3 mm from the back face, as illustrated in Figure 1, and were examined from the front face.

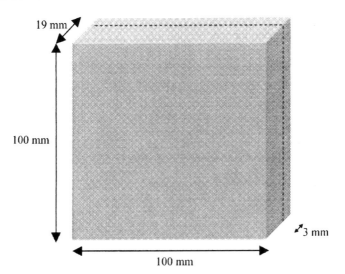

Figure 1. Illustration of seeded defect tiles.

RESULTS

In order to effectively evaluate the methodologies chosen for this investigation, tile "P" was selected for a more thorough analysis, and so comparisons will be made using this tile only. However, phased array data for all three tiles is presented below.

Conventional Ultrasonic Evaluation

A fast way to ultrasonically interrogate a volume of material for bulk defects using conventional transducers is to perform a pulse-echo C-Scan. Figure 2 shows a diagram of a SiC tile with flat-bottomed holes drilled in the back surface. Figure 3 shows a typical C-Scan of that SiC tile. The scan is oriented to correspond to the diagram shown in Fig. 2. The color scale represents relative response amplitude as shown in the accompanying scale. From such a scan, one can typically not only find the "defects" and obtain some measure of the relative "severity" of each defect (from the amplitude response on the left) but also determine their exact depth below the inspection surface (from the TOF response on the right).

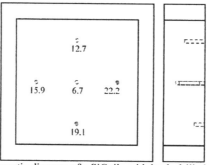

Figure 2. Schematic diagram of a SiC tile with back-drilled holes

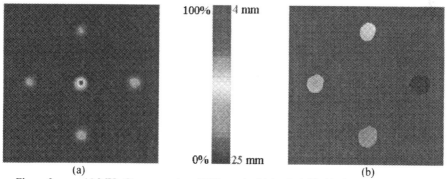

(a)

100%   4 mm

0%   25 mm

(b)

Figure 3.     10 MHz C-scan results of SiC panel with back-drilled holes using unfocused transducer. a) Peak amplitude and b) Time-Of-Flight (TOF) response. Percentage scale is for (a) and represent % of full-scale amplitude. Depth scale is for (b) and represents depth below front surface.

Figure 4 shows the conventional focused 5 MHz C-scan results for tile "P" when the transducer is focused on the top surface of the tile rather than on the defect plane. Only two "defects" (which are presumably regions of high concentration of the inclusion material) are evident (indicated by the arrows), along with a distinct band along the right edge of the tile. Note that the color scheme for defect amplitude is the same as shown in Figure 3a.

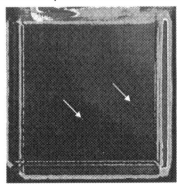

Figure 4. C-scan image of defect tile "P" acquired with conventional focused transducer focused on the tile's top surface.

Figure 5 shows the same tile acquired using the same transducer but now focused at the defect plane. Figure 5a shows the amplitude data while Figure 5b shows the TOF information. Because the position of the defect plane is not exact (and there are some minute variations in wave-speed in the intervening material) the C-scan actually records the amplitude and arrival of the peak response within a 1.5 mm window around the defect plane. When no defect is present in the defect layer the "peak" selected is the point of highest noise. Although the amplitude scan shows little variation in the background region, the TOF data is extremely noisy, as the time of arrival of this noise peak is somewhat random and can occur anywhere throughout the observed window. Thus, while some of the defect regions are barely detectable in the amplitude information they are more clearly discerned as "low noise" regions in the TOF data, and are therefore identified in both figures. The total scan time for this analysis was 60 minutes.

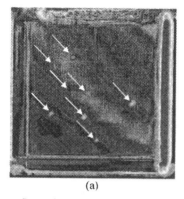 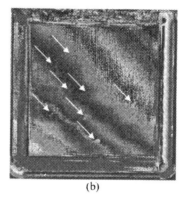

(a)                                    (b)

Figure 5. C-scan image of defect tile "P" acquired with conventional focused transducer focused on the defect plane showing a) amplitude and b) TOF data.

To overcome the effects of fixturing in Figure 5 (as evidenced by the noise at the top of the figure) the tile was rotated and rescanned. Two additional defects were revealed, yielding a total defect map shown in Figure 6. There are 10 localized concentrations of inclusions observed in the defect plane, in addition to the distinct band along the right-hand edge.

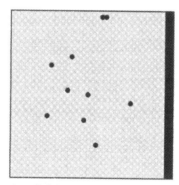

Figure 6. Schematic diagram showing all defects observed using conventional focused ultrasonic inspection.

Phased Array Ultrasonic Evaluation

The same block was then scanned with the phased array system and the results of this scan are shown in Figure 7. Here the same defects are observed as seen in the conventional ultrasound C-scan. In addition, owing to the reduced background noise (seen in the image as vertical striations associated with the "stitching" together of adjacent passes of the phased array) there are additional defects detected which were at or below the detection threshold with the conventional system. The total scan time for this analysis was approximately 3 minutes. However, as with the previous conventional analysis, the inspection time was significantly reduced by scanning only on a single plane (as the location of the defects was provided a priori).

A more realistic approach would be to scan the entire volume of the tile so that the improved sensitivity of phased array inspection could be applied independent of the defect depth.

For reference, the "A" and "V" tiles were also scanned under these conditions and the results of these scans are shown in Figure 8. Note that both of these scans indicate defect bands along the top and right edges of the tiles, even when the amplitude gain has been markedly increased (relative to the settings used on the "P" tile) so that the noise level is significantly more pronounced. Only the "V" tile shows a region of defect concentration similar to those observed in the "P" tile (in the lower right had corner).

Figure 7. C-scan image of defect tile "P" acquired with phased array transducer focused on the defect plane.

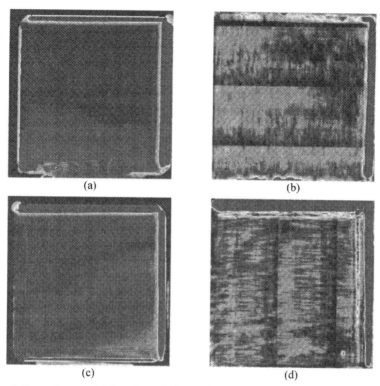

Figure 8. C-scan image of defect tiles a) "A" (acquired with conventional ultrasound), b) "A" (acquired with phased array system), c) "V" (acquired with conventional ultrasound) and d) "V" (acquired with phased array system), all focused on the defect plane.

The tile was therefore re-inspected using a depth-scanning modification in which the focus of the phased array transducer is scanned through the depth of the tile at each point in the raster scan. Because the depth of focus of the transducer is significantly larger than the desired lateral resolution the tile could be fully inspected by dividing the thickness into only 10 increments, so that the actual scan time of approximately 40 minutes was still less than the 60 minutes required using the conventional system at a single focus. The results of this true volumetric scan are shown in Figure 9. While the defects are again observed in the same locations shown in the earlier analyses there are two noticeable changes in the results. First, by tuning the focus of the transducer to the various depths, unwanted noise elements (such as the lingering response from the front surface reflection) are more efficiently excluded from the final result, yielding a dramatic reduction in the background "noise". Second, the dynamic range of the detection appears to have improved as well, as evidenced by the wider range of responses among the various defects detected. Both of these effects work to produce an improvement in the SNR for the defects. These results are shown in Figure 10 below.

Figure 9. C-scan image of defect tile "P" acquired with phased array transducer whose focus was scanned through the depth of the tile.

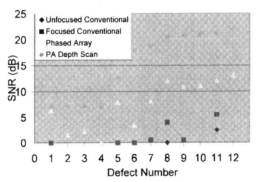

Figure 10. Comparison of Signal-to-Noise Ratio for tile "P" using various ultrasonic inspection methodologies.

DISCUSSION

While the true SNR would be best measured by determining a noise level from depths specifically not on the defect plane (i.e., those signals arriving before or after the defect reflection), the accumulation of this information for all locations in the image was too cumbersome to be effective. However, a reasonable comparison of the responses could be made by taking the peak of the response in the "defect-free" regions. Although this yields a significantly higher noise level, (as it assumes that the defect plane only contains inclusions at the indicated locations, which is not necessarily a valid assumption) it does give a lower limit to the effective SNR for each method. Therefore, in all four inspection methods the noise floor was determined as the three-sigma upper limit of the background (i.e., three standard deviations above the average of the "defect-free" regions). From the figure it is clear that the depth-scanned phased array method shows the largest SNR of the four. It is ~20 dB higher than the surface-focused conventional ultrasound, 15-20 dB higher than the focused conventional ultrasound, and 5-10 dB higher than the phased array system focused only on the defect plane.

## CONCLUSIONS AND FUTURE WORK

Ultrasonic methods have been evaluated for their potential for detecting "real" defects which may compromise ballistic performance in ceramic armor. Phased array inspection has been demonstrated to significantly improve upon the SNR of conventional ultrasonic inspection. If inspection speed is of primary interest, the phased array method can achieve improvements of between 5 and 10 dB, even at non-optimum frequencies, while reducing scan time 20-fold. However, if maximum sensitivity is the target the phased array system can improve SNR by as much as 20 dB while still reducing scan time by 35%.

Future efforts will be focused on maximizing SNR without compromising inspection time by evaluating dynamic focusing techniques. Here the sound is transmitted into the tile at a fixed, long depth-of-focus condition, so that only a single pulse of acoustic energy is required at each point. However, by electronically re-combining the received signals from the various array elements using different time delays between the elements several different effective focal lengths can be obtained from a single pulse-echo sequence. Thus, much of the SNR benefit achieved through volumetric scanning can be realized without a significant increase in overall scan time by performing a "virtual depth scan" in the post-inspection processing and analysis.

## REFERENCES

[1] J.M. Wells, W.H. Green and N. L. Rupert—""On the Visualization of Impact Damage in Armor Ceramics", Eng. Sci. and Eng. Proc. Vol, 22, Issue 3, H.T. Lin and M. Singh, eds, pgs 221-230, 2002

[2] Y. Tanabe, T. Saitoh, O. Wada, H. Tamura, and A. B. Sawaoka, "' An Overview of Impact Damages in Ceramic materials---For Impact below 2km/s" in Review of the Research Laboratory of Engineering materials, Tokyo Institute of Technology 19, 1994

[3] J.S. Steckenrider, W.A. Ellingson, J.M. Wheeler "Ultrasonic Techniques for Evaluation of SiC Armor Tile"in Ceramic Engineering and Science Proceedings, Volume 26, Issue 7, pgs 215-222. 2005

[4] R. Brennan, R. Haber, D. Niesz, J. McCauley "Non-Destructive Evaluation (NDE) of Ceramic Armor : Testing" in Ceramic Engineering and Science Proceedings, Volume 26, Issue 7, pgs 231-238. 2005

[5] G.P. Singh and J. W. Davies, "Multiple Transducer Ultrasonic Techniques: Phased Arrays" In Nondestructive Testing Handbook, 2nd Ed. , Volume 7, pgs284-297. 1991

[6] D. Lines, J. Skramstad, and R. Smith, " Rapid, Low-Cost, Full-Wave Form Mapping and Analysis with Ultrasonic Arrays", in Proc, 16th World Conference on Nondestructive Testing, September , 2004.

[7] J. Poguet and P. Ciorau, " Reproducibility and Reliability of NDT Phased Array Probes", in Proc, 16th World Conference on Nondestructive Testing, September , 2004.

# SPHERICAL INDENTATION OF SiC[1]

A. A. Wereszczak and K. E. Johanns
Ceramic Science and Technology
Oak Ridge National Laboratory
Oak Ridge, TN 37831

## ABSTRACT

Instrumented Hertzian indentation testing was performed on several grades of SiCs and the results and preliminary interpretations are presented. The grades included hot-pressed and sintered compositions. One of the hot-pressed grades was additionally subjected to high temperature heat treatment to produce a coarsened grain microstructure to enable the examination of exaggerated grain size on indentation response. Diamond spherical indenters were used in the testing. Indentation load, indentation depth of penetration, and acoustic activity were continually measured during each indentation test. Indentation response and postmortem analysis of induced damage (e.g., ring/cone, radial and median cracking, quasi-plasticity) are compared as a function of grain size. For the case of SiC-N, the instrumented spherical indentation showed that yielding initiated at an average contact stress 12-13 GPa and that there was another event (i.e., a noticeable rate increase in compliance probably associated with extensive ring and radial crack formations) occurring around an estimated average contact stress of 19 GPa.

## INTRODUCTION

There is strong interest to develop and refine Hertzian indentation test techniques to analyze armor ceramics because (1) there is much commonality of the damage produced from a static Hertzian indent and damage produced from a ballistic impact and (2) appropriately chosen indenter diameters can generate representative contact stresses that can occur in impact. Another advantage of spherical indentation over pyramidal-based indentation is the entire evolution of damage modes can be produced as progression occurs from elasticity to full quasi-plasticity. Prospects exist to quantify mechanical parameters that can be used to rank contact-damage resistance in armor ceramics and that perhaps can be ideally used as input in ballistic impact damage models. Instrumented (i.e., load and indenter depth of penetration are independently measured during a controlled load-unload force profile) Hertzian indentation can be simple, quick, and the sample preparation and postmortem analyses are also relatively straightforward and inexpensive. The concept of instrumentation has been around for at least 15 years now [1-2]. The very fact that it can generate similar (and controlled) damage and stress magnitudes of interest, and facilitate its systematic characterization, creates rationale to exploit it.

Literature is plentiful involving Hertzian indentation of ceramics [1-3]; however, the desire to examine indentation responses of "bulk" armor ceramics creates some unique challenges. For example, the hardness of almost all candidate ceramic armors is higher than that of WC/Co ball indenters (the most widely-used indenter in the literature) so the use of diamond indenters is

[1] Research sponsored by WFO sponsor US Army Tank-Automotive Research, Development, and Engineering Center under contract DE-AC05-00OR22725 with UT-Battelle, LLC.

desirable. However, diamond spherical indenters are not common and the user must rely on their custom (i.e., expensive) manufacturing. A compromise must be struck in that the indenter diameter must be large enough to indent a large area (i.e., sample a bulk material response) yet be small enough so that their manufacturing does not become cost prohibitive or require large compressive forces that could damage them.

This article describes an instrumented indentation test system that was developed to test armor ceramics and use it to compare Hertzian indentation responses of six silicon carbides (SiC). It is demonstrated that subtle (and not so subtle) differences in high-stress-induced damage mechanism activity and evolution can be identified.

PROCEDURES
Indenter System and Indenters

The instrumented indentation test system is shown in Fig. 1. A computer-controlled X and Y stage translate a specimen between the focal point under an optical microscope and an indenter mounted at the bottom of a computer-controlled stepper-motor Z stage. A load cell is in the load train and the three capacitance gages measure indenter displacement during indentation loading and unloading. The three displacements measured with the capacitance gages are averaged in later data reduction involving graphical analysis of load-displacement curves. The test system also includes the measurement of acoustic emission (sensor typically glued to a specimen) whose data are routinely analyzed as a function of load (during loading or unloading). The system contains two digital video cameras - one is mounted on the optical microscope (not shown in Fig. 1) and the other is used to watch the indenter approach the sample surface. The user identifies coordinates through the optical microscope for the desired locations of the indents, the X-Y stage accurately translates the specimen under the indenter, and then the user applies a controlled triangular waveform for displacement-controlled loading up to a desired maximum load followed by displacement controlled unloading. Typically used displacement rates are 0.1 and 1.0 μm/s.

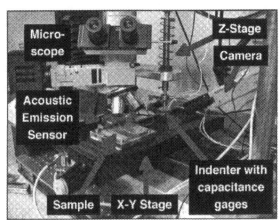

Figure 1. Instrumented indentation setup.

Custom-made diamond spherical indenters[2] are used for the testing. Two primary lessons have been learned over the last few years since this work was initiated [6]. First, an indenter diameter size in the range of 250 to 1000 microns enables a desirable (i.e., bulk) area and volume of test material to be sampled and enables the generation of many tens of GPa of pressure at sufficiently low compressive loads that do not cause indenter damage. This effect is illustrated in Figs. 2-3. Expressions for the determined curves in Fig. 3 can be found in the classical Johnson text [7]. Tens of GPa of pressure can be produced through loading only to 100-200N (versus the need to apply tens of kN with a 6350 μm or 6.35 mm diameter ball); a relatively low force that allows for long indenter lifetime. Additionally, the contact area diameter produced with indenter diameters of 250-1000 μm is many tens of microns (i.e., at least an order of magnitude greater than the typical average grain sizes of the tested ceramics), so confidence exists that a bulk response is sampled (unlike what would be produced with a "nanoindenter"). The other primary learned lesson is surface roughness of the indenters need to be minimized in order to measure repetitive responses and to generate "clean" looking indents. An example of the high quality surface condition on one of our indenters is shown in Fig. 4. A disadvantage of these small sized indenters is they often are applying at least 10 GPa of average contact stress at just a few Newtons of force, so attempting to quantify the load associated with yield initiation can be elusive. Our present experimental strategy to identify yield is to use our largest diameter (1000 μm) because a higher force can be applied before that initiates. As will be seen, the average contact stress of yielding in SiC-N was approximately 12-13 GPa - that can be readily identified with a 1000 μm indenter at around 80-90N whereas that contact stress is achieved with less than 10 N with a 500 μm diameter or smaller.

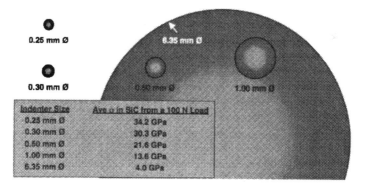

Figure 2. Comparison of drawn-to-scale indent diameter size and the average contact stress that each experience when loaded to 100 N onto SiC (E = 450 GPa, ν = 0.18). The four smaller sizes are used in these studies.

---

[2] Gilmore Diamond Tool, Attleboro, MA.

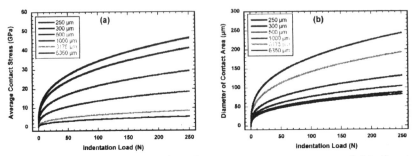

Figure 3. Hertz elastic contact functions for (a) average contact stress and (b) diameter of contact area as a function of indentation load and indenter diameter when a diamond indenter is in contact with SiC. Simulation uses E = 1141 GPa and $\nu$ = 0.07 for diamond and E = 450 GPa and $\nu$ = 0.18 for SiC.

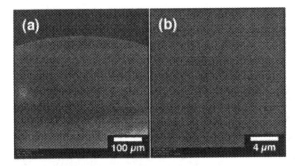

Figure 4. Representative surface condition of our custom made diamond indenters manufactured using the "Precision Spindle Method" (Gilmore Diamond Tools, Attleboro, MA). For this example, low (a) and high (b) magnifications of a 1000 μm diameter indenter are shown.

Evaluated Silicon Carbides (SiCs)

Six SiCs were evaluated for this study and information on each is provided in Table I. Four (SiC-N, SiC-B, SiC-SC-1R, and Ceralloy 146) were made by hot pressing and were fully dense. The SiC-SC-1R serves as a useful model material because its grain size is 2-3 times smaller than that for the SiC-N, SiC-B, and Ceralloy 146. The fifth evaluated material was SiC-N that was additionally subjected to a 2500°C heat treatment [8] with the intent to promote grain growth without compromise to hardness, and enable a preliminary examination of grain size effect on indentation response. The sixth SiC was Hexoloy SA and is made by pressureless sintering, and unlike the hot pressed compositions, had a few percent of bulk porosity.

Table I.  SiC grades evaluated and elastic properties used in analysis.

| SiC Grade | Manufacturer | Young's Modulus (GPa) | Poisson's Ratio |
|---|---|---|---|
| SiC-N | Cercom | 450 | 0.17 |
| SiC-N (2500)* | Cercom | 450 | 0.17 |
| SiC-B | Cercom | 450 | 0.17 |
| SiC-SC-1R | Cercom | 450 | 0.17 |
| Ceralloy 146 | Ceradyne | 450 | 0.17 |
| Hexoloy SA | Saint-Gobain/Carborundum | 430 | 0.18 |

Indentation Procedures

The SiC-N was subjected to the most indentation testing.  250, 300, 500, and 1000 μm diameter indenters were used.  Load-indent-depth-of-penetration curves were measured in all tests and acoustic emission was monitored as well.  A displacement rate of 1.0 μm/s was used up to a predetermined maximum load and back to zero force.  The 300 μm diameter was also used to perform cyclic testing to help interpret hysteresis for an undamaged and damaged state of material.

The indentation responses of the other five SiCs were primarily evaluated using a 300 μm diameter indenter.  Maximum indentation loads were the same among them and the same as those for the SiC-N to enable comparison of the responses.

Analysis

Load-unload curves were compared for each of the materials.  Additionally, indents were examined with differential interference contrast (DIC or "Nomarski") imaging.

Average Hertzian contact stresses are reported below for the sake of discussion.  The maximum stress is actually 50% higher than that value and that stress location is under the indenter at its contact centerline.  For damage initiation under the indenter, that damage could undergo initiation because of that maximum value and not the reported average Hertzian stress.  Once damage initiates, the Hertzian contact stress distributions cannot be calculated from the classical elastic equations.

RESULTS

Indentation of the SiC-N using 250, 300, 500, and 1000 μm indenters showed that the level of produced damage was equivalent for the same average contact stress but that the indent size was larger with larger indenter diameter.  This is illustrated in Fig. 5.  The advantage of the use of the 1000 μm indenter is an appreciable contact force can be applied before apparent yielding is initiated.  In this case, that force was at approximately 80-90N corresponding to an average contact stress of ~12 GPa (a value equivalent to the Hugoniot Elastic Limit for hot pressed SiCs [9-10]).  That average contact stress was quickly eclipsed during the indentation loadings of the 250, 300, and 500 μm diamond indenters.  For this reason, we favor the use of the

1000 μm diameter indenter to estimate yield stress. The remaining imaged indents will be from testing with the 300 μm diameter indenter - that indenter size was effective at producing estimated average contact stresses exceeding 20 GPa.

The damage under the indent (designated as the "quasi-plastic" damage zone) has been studied extensive by Lawn [3] and others. The quasi-plastic zone resembles the plastic zones that occur in metals; however, it consists of closed Mode II cracks with internal sliding friction (shear faulting) occurring at the weak planes. The interface bonded specimen (IBS) has been frequently used to study this quasi-plastic zone; however, even though it is an elegant way to explore a version of subsurface damage from spherical indentation, its usage when attempting to quantify indentation response and parameters is fraught with peril and is therefore not used here. The two free surfaces in the IBS do not accommodate the large shear stresses that exist when indenting a bulk specimen which in turn is likely to produce different extents of damage mechanisms or different damage mechanisms altogether. The important median crack system will likely never be produced with an IBS - if it is not produced like it can be when a bulk specimen is spherically loaded, then what effect does its absence have on the evolution of the other crack systems that do form with the IBS. The usage of a focused ion beam (FIB) miller has recently been employed for examining subsurface crack formation and structural characteristics [11] but that was with a nanoindenter-sized indent (~ few microns). The authors are in the process of performing such FIB'ing with one of our Hertzian indents in SiC-N as well as other post-indentation sectioning methods.

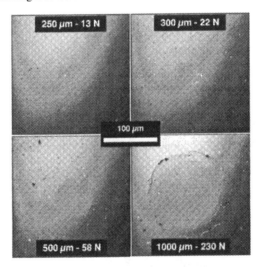

Figure 5. Indentation response in SiC-N with four different diameters to a maximum load (diameter-load pair shown in each image) that produced an (idealized) average contact stress of 18 GPa. Yielding had obviously been exceeded at this contact stress.

Indenter size and damage as a function of force is shown in Fig. 6 for SiC-N. The 13, 22, 36, 58, 80, and 111N forces, if only elasticity were present, correspond to an average contact stress of 15.4, 18.3, 21.6, 25.3, 28.2, 31.4 GPa, respectively. Obviously the elastic limit was exceeded at all these forces so the authors fully realize that those reported average contact stresses are inaccurate; however, it is convenient for the sake of discussion to report those values nonetheless. The 13N force shows the disadvantage of the use of the 300 µm diameter indenter to identify yield; the average contact stress was already above 15 GPa. Concentric ring cracking initiated between 22 and 36 N (18.3 and 21.6 GPa) and median cracking initiates between 36 and 58 N (21.6 and 25.3 GPa) and they both became more extensive above 58 N (25.3 GPa). These ranges of stress are similar to the threshold stress level that Holmquist has modeled and identified for overcoming dwell [12]. For the present, the cracks radially oriented are identified as "radial" cracks; however, without sectioning of the volume under the indent, the authors cannot discount the possibility that some of these cracks could be median cracks that have advanced to the surface. A nice survey of the various crack types can be found in Cook and Pharr [1].

An effective way to portray the extent of the indent damage in Fig. 6 is through cyclic indentation (results shown in Fig. 6). The amount of hysteresis as a function of maximum load is linked to the amount of accrued indent damage. For example, there is only a small amount of apparent yielding at a peak load of 13 N in Fig. 5 and that bears out as little or no hysteresis in the corresponding graph in Fig. 6. As higher peak loads are applied, the amount of hysteresis continues to increase. The amount of hysteresis becomes noticeable at 36 N (21.6 GPa) when concentric ring cracking has initiated and starts to increase at a faster rate once radial (or median or both) cracking initiates between 36 and 58 N (21.6 and 25.3 GPa), and becomes more pronounced at higher forces and average contact stresses. The cyclic tests also suggest that additional damage does not significantly accumulate when reloaded to 80 N (28.2 GPa) but that it incrementally increases when reloaded to 111 N (31.4 GPa).

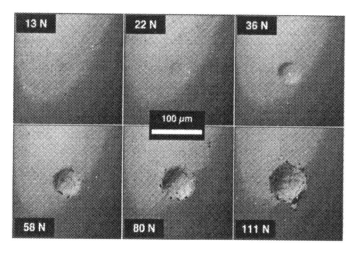

Figure 6.    Indentation response in SiC-N as a function of load with a 300 µm diameter indenter.

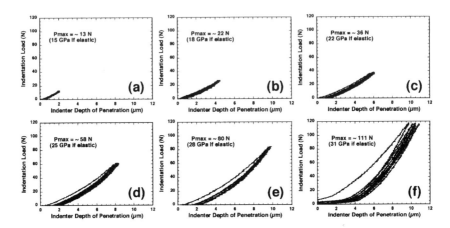

Figure 7.  Cyclic load-indenter-depth-of-penetration as a function of peak load in SiC-N with a 300 μm diameter indenter.  The six graphs indirectly correspond to the indents shown in Fig. 5; however, those images in Fig. 5 are from a single cycle indent sequence.

Indent images as a function of peak force for SiC-B, SiC-N heat treated at 2500°C, SiC-SC-1R, Ceralloy 146, and Hexoloy SA are respectively shown in Figs. 8-12. The SiC-B had a similar response to the SiC-N. The coarsened microstructure of the 2500°C heat-treated SiC-N shows extensive cracking about the large grains. Generated indents in the SiC-SC-1R looked ideal ostensibly because of the material's fine grain size. Indentation response of the Ceralloy 146 looks to be equivalent to the SiC-N and SiC-B though there is a suggestion that radial cracking started at a lower load than the SiC-N and SiC-B. Extensive amounts of concentric ring and radial cracking occurred in the Hexoloy SA; the greater amount of damage is likely a consequence of the inherent porosity that it has (and that the hot pressed SiCs do not).

A comparison of the indents generated at the maximum load of 111N for the six SiCs are shown in Fig. 13. Cracking is extensive in each with some of the grades have more than others. Even though the heat treated SiC-N has very large grains and the fracturing inside its indent appears to be severe, the extent of radial cracking is arguably no more than in the other materials. The Hexoloy SA has more concentric ring cracking than the other SiCs. The load-penetration curves in Fig. 13 show that the indent depth of penetration for the SiC-SC-1R and the heat treated SiC-N and the SiC is more than for the other SiCs - the former appears to have accommodated that whereas the extensive amount fracturing in the latter's indent is consistent with that deep indenter penetration.

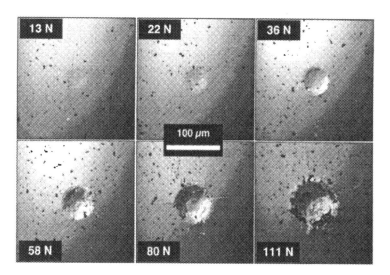

Figure 8.   Indentation response in SiC-B as a function of load with a 300 μm diameter indenter.

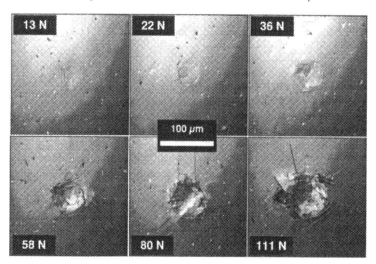

Figure 9.   Indentation response in 2500°C heat treated SiC-N as a function of load with a 300 μm diameter indenter.

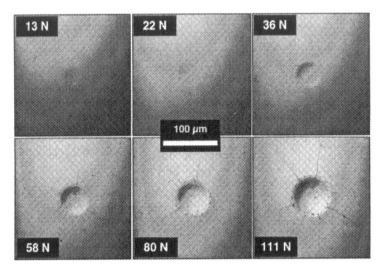

Figure 10. Indentation response in SiC-SC-1R as a function of load with a 300 μm diameter indenter.

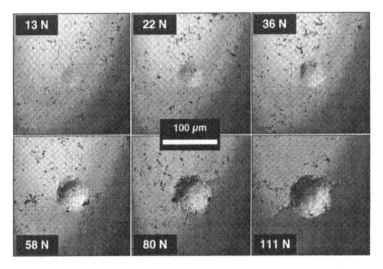

Figure 11. Indentation response in Ceralloy 146 as a function of load with a 300 μm diameter indenter.

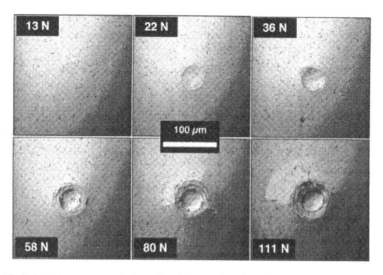

Figure 12. Indentation response in Hexoloy SA as a function of load with a 300 μm diameter indenter.

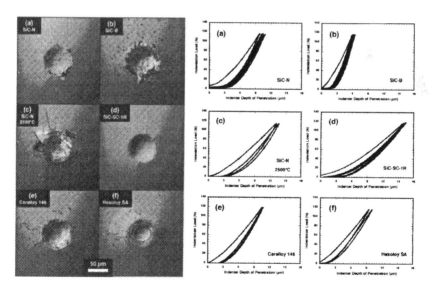

Figure 13. Comparison of Hertzian indents generated with a 300 μm diameter indenter and a maximum load of 111 N and companion comparison of load - penetration curves.

One anecdotal (?) observation can be made regarding the metallographically prepared surfaces of the six SiCs shown in Fig. 13. All these specimens were prepared and polished under identical conditions yet there is an appreciable amount of grain pullout in the SiC-B and Ceralloy 146. The fraction of apparent surface porosity in the Hexoloy SA is undoubtedly contributed to by bulk porosity inherent to the material. It is outside the scope of the present study to study these observations in depth; however, they do suggest that the grain boundary strength in the SiC-B and Ceralloy 146 may not be as high as it is in the SiC-N and SiC-SC-1R (which exhibit little or no grain pullout from the metallography). The authors do raise this observation though because such grain boundary strength would seem to be influential in indentation response and overall contact-damage evolution.

Acoustic emission histories for the Fig. 12 indents are shown in Fig. 14. For a 45 db threshold, no activity was detected when the SiC-N, SiC-B, and Ceralloy 146 was indented with a 300 μm indenter. A lot of acoustic activity occurred in the heat treated SiC-N and Hexoloy SA.

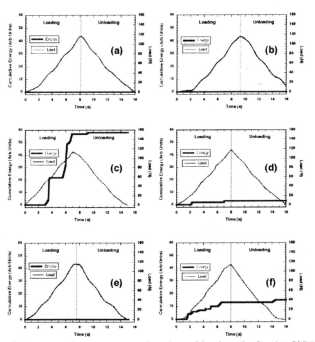

Figure 14. Cumulative acoustic energy as a function of load cycle for (a) SiC-N, (b) SiC-B, (c) SiC-N 2500° treatment, (d) SiC-SC-1R, (e) Ceralloy 146, and (f) Hexoloy SA.

The hysteresis of the load-penetration curves was studied in more detail for the four hot pressed grades. The load-penetration curves for each are shown in Fig. 15. Load to the two-thirds power is plotted as an abscissa because the depth of penetration as a function of $P^{2/3}$ is linear when pure elasticity exists and it can be a convenient way to portray this functionality for spherical indentation. The shapes of the curves for the SiC-N, SiC-B and Ceralloy 146 are similar with subtle differences in their hystereses. The SiC-SC-1R is noticeably different. If one subtracts the two curves (see Fig. 15e), then some of the subtle differences in Figs. 15a-d become more pronounced. The slopes of the curves in Fig. 15e are continually decreasing for the SiC-N, SiC-B, and Ceralloy 146 inferring an apparent "hardening" response whereas the slope of the curve for the SiC-SC-1R is increasing. Load-penetration data was fitted to Field and Swain's model [5] but no combination of adjustable parameters could satisfactorily fit the data. Efforts are underway to develop a model that can better simulate our experimental data.

Lastly, the loads and other dependent parameters discussed in this work are undergoing interpretation for linkability to impact threshold velocities. Lundberg [13] recently attempted to link that to fracture toughness, but the authors will explore any associations with the observations overviewed in the study.

## SUMMARY

Instrumented Hertzian indentation of ceramics enables the engineer or scientist to assess contact damage responsiveness. It can be used to quantify the onset of yield and show changes in rates of compliance increase that is likely do to extensive ring and radial cracking. Similar indentation responses were observed with SiC-N, SiC-B, and Ceralloy 146, though there was anecdotal evidence to suggest that the grain boundary strength the latter two was less than that for the SiC-N. The average contact stress for yield initiation in SiC-N was ~ 12-13 GPa, and at around 19 GPa, the rate of compliance increase increased consistent with extensive concentric ring and radial cracking. SiC-SC-1R was a useful model material for indentation analysis because of its fine grain size. 2500°C heat treated SiC-N showed a small amount of greater indenter depth of penetration than SiC-N, SiC-B, and Ceralloy 146. Hexoloy SA was less resistant to contact damage than the other materials - most likely because of the few percent of inherent porosity that it has. Cyclic indentation results and the traditional approach to subtract the loading and unloading penetration curves from one another aids in the interpretation of Hertzian contact damage of armor ceramics.

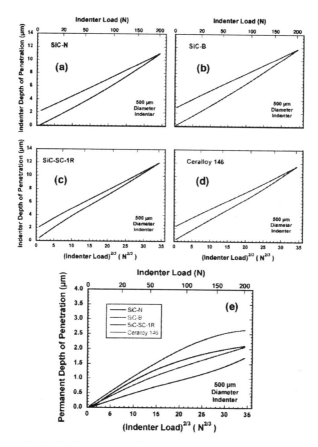

Figure 15. One cycle load-indenter penetration curves for (a) SiC-N, (b) SiC-B, (c) SiC-SC-1R, and (d) Ceralloy 146 indented with a 500 μm indenter. Hysteresis of each is represented by the permanent depth of penetration curves in (e).

## ACKNOWLEDGEMENTS

The authors wish to thank ORNL's J. E. Shelton for specimen preparation and T. Geer for the metallography, Ceramatec's M. Ferber for the use of the data analysis software, US Army TARDEC's L. Franks and Ceradyne's M. Normandia for supplying some of the tested SiCs (latter supplied materials while employed at the US ARL), US ARL's J. Swab and J. LaSalvia for their helpful input, and US ARL's G. Gilde for the SiC-N heat treating.

## REFERENCES

[1]  R. F. Cook and G. M. Pharr, "Direct Observation and Analysis of Indentation Cracking in Glasses and Ceramics," *J. Am. Ceram. Soc.*, **73**, 787-817 (1990).

[2]  W. C. Oliver and G. M. Pharr, "An Improved Technique for Determining Hardness and Elastic Modulus Using Load and Displacement Sensing Indentation Experiments," J. Mater. Res., **7**, 1564-83 (1992).

[3]  B. R. Lawn, "Indentation of Ceramics with Spheres:  A Century after Hertz," *J. Am. Ceram. Soc.*, **81**, 1977-94 (1998).

[4]  A. C. Fischer-Cripps and B. R. Lawn, "Stress Analysis of Contact Deformation in Quasi-Plastic Ceramics," *J. Am. Ceram. Soc.*, **79**, 2609-18 (1996).

[5]  J. S. Field and M. V. Swain, "A Simple Predictive Model for Spherical Indentation," *J. Mater. Res.*, **8**, 297-306 (1993).

[6]  A. A. Wereszczak and R. H. Kraft, "Instrumented Hertzian Indentation of Armor Ceramics," *Cer. Engrg. Sci. Proc.*, **23**, 53-64 (2002).

[7]  K. L. Johnson, *Contact Mechanics*, Cambridge University Press, Cambridge, United Kingdom, 1985.

[8]  A. A. Wereszczak, H. -T. Lin, and G. A. Gilde, "The Effect of Grain Growth on Hardness in Hot-Pressed Silicon Carbide," in review, *J. Mat. Sci.*, 2005.

[9]  R. Feng, G. F. Raiser, and Y. M. Gupta, "Shock Response of Polycrystalline Silicon Carbide Undergoing Inelastic Deformation," *J. Appl. Phys.*, **79**, 1378-87 (1996).

[10]  D. E. Grady, "Shock-Wave Compression of Brittle Solids," *Mech. Mat.*, **29**, 181-203 (1998).

[11]  Z. -H. Xie, M. Hoffman, R. J. Moon, P. R. Munroe, and Y. -B. Cheng, "Subsurface Indentation Damage and Mechanical Characterization of $\alpha$-Sialon Ceramics," *J. Am. Ceram. Soc.*, **87**, 2114-24 (2004).

[12]  T. J. Holmquist, C. E. Anderson, Jr., and T. Behner, "Design, Analysis and Testing of an Unconfined Ceramic Target to Induce Dwell," given at 22nd International Symposium on Ballistics, Vancouver, BC, Canada, 14-18 November 2005.

[13]  P. Lundberg and B. Lundberg, "Transition Between Interface Defeat and Penetration for Tungsten Projectiles and Four Silicon Carbide Materials," *Int. J. Impact Engrg.*, **31**, 781-92 (2005).

# DAMAGE MODES CORRELATED TO THE DYNAMIC RESPONSE OF SIC-N

H. Luo
School of Mechanical and Aerospace Engineering
Oklahoma State University
218 Engineering North, Stillwater, OK 74078

W. Chen
Schools of Aero/Astro. and Materials Engineering
Purdue University
315 N. Grant St., West Lafayette, IN 47907

ABSTRACT

The damage modes in a hot-pressed silicon carbide (SiC-N) specimen have been correlated to its dynamic compressive response at high strain rates. We employed a novel dynamic loading/reloading experimental technique modified from a split Hopkinson pressure bar (SHPB) to determine the dynamic properties and to record the damage/failure modes in the ceramic specimen, in which a ceramic specimen was loaded by two consecutive stress pulses. The first pulse determines the dynamic response of the intact ceramic material and then crushes the specimen to a desired damage level. The second pulse then determines the dynamic compressive constitutive behavior of the damaged but still interlocked ceramic specimen. The first pulses were varied slightly to control the damage levels in the ceramic specimen while the second pulse was maintained identical. The results show that the compressive strengths of damaged ceramics depend on a critical level of damage, below which the specimen retains its axial load-bearing capacity and only axial cracks are observed in the specimen. When the specimen is critically damaged, axial cracks and isolated pulverized regions are observed. When the specimen is damaged beyond the critical level, the ceramic specimen is crushed into cracked particles with pulverized (comminuted) materials along the particle boundaries, which displays a granular flow behavior in its stress-strain curve.

## INTRODUCTION

Due to a favorable combination of mechanical and physical properties, ceramic materials have been increasingly applied in armor systems that are subjected to high pressures under impact loading. When a projectile penetrates into a ceramic target, a comminuted zone is created ahead of the penetrator, which is accompanied by the dynamic propagation of cracks into the target material. The comminuted ceramic is in direct contact with the penetrator and erodes the penetrator. Under confinement by surrounding cracked ceramic and intact ceramic in the far field, the comminuted ceramic exhibits some strength. This strength of the failed ceramic is found to be dependent on the confining pressure.[1-5] Extensive research efforts have been invested in the understanding of impact response of intact ceramic materials or ceramic powder using tools such as plate impact, modified SHPB, and projectile penetration.[6-11] Recently, a novel two-pulse SHPB technique loads the ceramic specimen by two successive stress pulses to characterize the dynamic compressive behavior of alumina.[12,13] However, more experimental data on failed ceramics are desired to support computational simulations, which currently use many assumed strength characteristics.[14,15] Refined two-pulse SHPB experiments were conducted to investigate the dynamic compressive response of damaged SiC-N ceramic as a function of the strain rates, damage levels, volume dilatation and lateral confinements.[16,17] The mechanical behavior of the comminuted ceramic was not found to be very sensitive to the strain rates and volume dilatation once the damage was accumulated beyond a critical level. The critical damage may or may not be reached depending on the control over the profiles of the first pulse.

The critical damage is difficult to identify. It is attempted in this paper to relate the crack/damage patterns in the specimen to its dynamic mechanical response. Three different types of crack patterns are presented. The first is the crack pattern found in slightly damaged SiC-N specimens, then those found in intermediately (nearly critically) damaged specimens are presented. The last type is crack/damage pattern found in specimens heavily damaged by mechanical loading. Further SEM observations reveal more detailed microstructural failure mechanisms. After correlating the crack/damage patterns to the mechanical responses, a Mohr-Coulomb model and the JH-1 model for the damaged SiC-N are discussed and compared. The model parameters were determined from the experimental results.

CRACK PATTERNS OF DAMAGED SIC-N

After the mechanical experiments with a modified SHPB[16,17], the crack/damage patterns in specimens confined with steel sleeves of 0.8-mm wall thickness, which provide 104-MPa confinement for the failed ceramic inside, were imaged by a digital camera. Figures 1(a) and 1(b) show crack patterns in a slightly damaged SiC-N specimen as observed on end surface and from an inner axial section, respectively. The section was cut with a low-speed diamond saw after the specimen was loaded. Only axial cracks are observed in such slightly damaged (or damaged below the critical level) specimens. These specimens can carry further axial compressive stresses with minimum effects from the axial cracking, even though the cracks are clearly observable.

Figures 2(a) and 2(b) show the typical crack/damage patterns of a nearly critically damaged SiC-N specimen on an end surface and in an axial section, respectively. The crack patterns in these transitionally damaged specimens may be roughly divided into two areas: a heavily damaged area near the corner of the specimen and around large chunks of the cracked ceramic material, and slight damage in the rest of the specimen. This seems to indicate that crack/damage density can increase rapidly at and near stress-concentrated areas in the ceramic specimen. If mechanical load continues, the crack/damage density in these localized high-stressed regions can increase rapidly. The specimen starts to lose its axial loading bearing capacity as the density of the localized damage increases.

Figure 3(a) shows the axial crack patterns of a heavily damaged (damaged beyond critical level) SiC-N specimen on an axial section. The detailed local crack patterns in the axial section were obtained with a metallographic optical microscope. Figure 3(b) shows more details of the local crack patterns, also in the axial section in the heavily damaged SiC-N specimen. The detailed image shows that even within a large SiC-N chunk in the heavily damaged specimen, there are many inner cracks that divide the chunk further into many small fragments. It is anticipated that these fragments would further slide/crack upon further mechanical loading. Between large SiC-N chunks are many more but much smaller comminuted particles. The length direction of the SiC-N fragments is mainly along the cylindrical axis of specimen, i.e., the loading direction. The cracks observed in specimens at all damage levels reveal that the compressive load induced axial splitting results in many narrow long fragments along the loading axial. Then the narrow long fragments further split before fracturing into smaller fragments. Under further compressive loading, the interfacial region between the fragments was comminuted to further small rubbles. When the specimen is damaged to this stage, the corresponding stress-strain behavior of the sample becomes a granular flow like or perfectly plastic like response. The axial load bearing capacity of the sample depends directly on the confinement level in the radial directions.

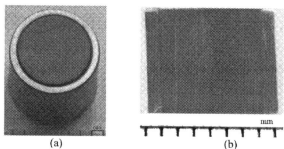

(a)                                          (b)

Figure 1. Crack pattern in a slightly damaged SiC-N specimen
(a) on an end surface; (b) in an axial section.

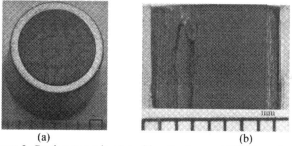

(a)                                          (b)

Figure 2. Crack patterns in a transitionally damaged SiC-N specimen
(a) on an end surface; (b) in an axial section.

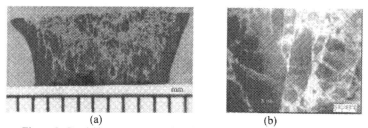

(a)                                          (b)

Figure 3. Crack/damage patterns in a heavily damaged SiC-N specimen
(a) in an axial section; (b) detailed crack/damage pattern.

MICROSTRUCTURAL OBSERVATION AND ANALYSIS OF SIC-N

The microstructural observations in SEM for the damaged SiC-N ceramic were conducted on specimens in each level of damage. The specimens were confined with 0.8-mm thick steel sleeves. For SEM observation, a cracked specimen was mounted on a sample holder. The sample rubble was typically grouped into large and small particle sizes. The samples on the holders are then sputtering coated, and then studied in SEM with high magnification. Figure 4(a) shows the end surface profile of an original SiC-N specimen without mechanical loading history, where no visible damage or crack is observed. Figure 4(b) shows the crack pattern on an end surface in a slightly damaged specimen, where surface cracks are clearly visible. However, there is no fine debris (comminuted ceramics) inside the cracks. Figure 5 (a) shows a large broken SiC-N chunk that has an inner crack. Figure 5(b) shows the details on the fracture surface of the large chunk, revealing intergranular fracture paths on the surface formed by axial splitting. Figure 6(a) shows a SEM micrograph of a specimen end surface of a transitionally damaged SiC-N. The image is focused on a locally transitional area between the less damaged fragments and heavily damaged comminuted rubble. This rubble zone expands as the sample deforms further. A detailed image of the fine rubble zone is shown in Fig. 6(b).

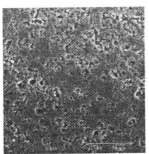 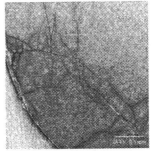

(a) (the scale is 50 μm)          (b) (the scale is 0.5 mm)

Figure 4. Crack patterns on the specimen end surface
(a) original SiC-N; (b) slightly damaged SiC-N.

 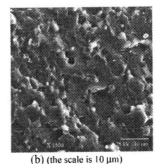

(a) (the scale is 0.25 mm)          (b) (the scale is 10 μm)

Figure 5. SEM micrographs of the damaged SiC-N
(a) a large chunk with in inner crack; (b) fracture surface of the large chunk.

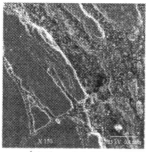 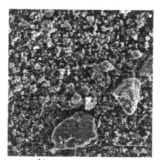

(a) (the scale is 0.1 mm)          (b) (the scale is 40 μm)
Figure 6. SEM micrograph of a nearly critically damaged SiC-N specimen
(a) transitionally damaged area; (b) details of the comminuted rubble.

The SEM observation indicates that intergranular fracture dominated the fracturing processes in the SiC-N specimen under compression. For the slightly damaged specimen, the fracture pattern is fragments with macro main cracks plus short and small sub-cracks. For the specimens damaged to nearly critical level by mechanical loading, heavily damaged regions with small particles start to fill in between large chunks of cracked ceramic material. The large fragments still remain slightly damaged. For the heavily damaged specimens, the large chunks are further fractured into smaller chunks. In fact, no large chunks are left in the specimen. Furthermore, these further fractured small chunks (particles) are surrounded by fine particles (comminuted ceramic) so that the large particles can slide against each other relatively easily, which facilitates granular flow.

FAILURE CRITERION AND DAMAGED MODELS FOR SIC-N
Once the mechanical response is determined and the failure mode characterized, it is desired to use an analytical model to describe the experimental results. The failure behavior of brittle materials is considered to obey the Mohr-Coulomb failure criterion under either quasi-static or dynamic loading conditions.[18] The pressure-dependent shear strength model has been used to describe the comminuted region in the ceramic target.[19-21] A number of compression failure models are also related to pressure-dependent shear failure such as the wing-crack or z-crack models.[22-24] Some material models are based on continuum damage theories in which the pure effect of fracture is homogenized as a degradation of the strength and elastic modulus of the materials.[25,26] The damage parameters were typically selected as the fragmentation size and density. The micro-mechanical models in terms of energy[27], micro-cracking evolution[28], analytical mechanics-based model[29] for dynamic fragmentation of ceramics, probabilistic-deterministic transition[30], damage and multi-scale model[31] were also developed. As a phenomenological model, Johnson-Holmquist (JH) model is well known and widely employed to describe the dynamic behaviors of brittle ceramics under shock and high strain rate loading conditions without considering into the details of crack growth.[32,33] Three closely related constitutive models for brittle materials, JH-1, JH-B and JH-2, were developed.[14,15,34-36] Modeling penetration processes in ceramic targets has recently been a focus of research. The acquired experimental results are intentionally used in computer simulations of ballistic events,

and thus need to be expressed in the forms of material models with constants determined. Here we attempt to use a model with a simple form to describe the failure behavior of SiC-N. The Mohr-Coulomb failure criterion is described as

$$|\tau| + \alpha p = \tau_0 \tag{1}$$

where $|\tau|$ is the absolute of the shear strength of the material, $p$ is the hydrostatic pressure, $\tau_0$ is the pure shear strength; and $\alpha$ is the internal friction coefficient. As an equivalent form in terms of the principal stresses, an analysis on the micro-sliding fracture mechanics[5,22-24] indicated that a 'wing crack' would start to propagate when,

$$\sigma_1 = c\sigma_3 - \sigma_c \tag{2}$$

where $\sigma_c$ is the unconfined compressive strength, and $c$ is material constant, the three principal stresses are $\sigma_1$, $\sigma_2$ and $\sigma_3$ ($\sigma_1 > \sigma_2 > \sigma_3$). In the case of a cylindrical specimen subjected to axially symmetric compression with lateral confinement, $\sigma_1$ is the axial compressive stress $\sigma_A$, $\sigma_3$ is the transverse or confining stress $\sigma_T$. Three principal stresses and the hydrostatic stress are thus given by

$$\sigma_1 = \sigma_A, \quad \sigma_2 = \sigma_3 = \sigma_T, \quad p = \tfrac{1}{3}\sigma_{kk} = \tfrac{1}{3}(\sigma_A + 2\sigma_T) \tag{3}$$

In terms of three principal stresses, the Von Mises equivalent stress is:

$$\sigma_e = \sqrt{\tfrac{1}{2}[(\sigma_1 - \sigma_2)^2 + (\sigma_2 - \sigma_3)^2 + (\sigma_3 - \sigma_1)^2]} = \sigma_A - \sigma_T \tag{4}$$

At this stress level, the intact brittle material will fail under the tri-axial compression. In terms of principal stresses, the shear strength is given by:

$$\tau = \tfrac{1}{2}(\sigma_1 - \sigma_2) = \tfrac{1}{2}(\sigma_A - \sigma_T) \tag{5}$$

i.e., the shear strength is half of the equivalent stress in this special case.

With equations (3), (4) and (5), the shear strength-pressure relation is related to the experimentally measurable quantities, compressive strength and lateral confining pressure. The Mohr-Coulomb model is then employed to describe the strength of the damaged SiC-N. The actual results from experiments were published earlier.[16] Parameters in the model are determined from the published results and shown in Table I. The graphic form is the same as that for JH-1 presented next.

Table I. Model parameters for the damaged SiC-N

| JH-1 model | Mohr-Coulomb model | | | |
|---|---|---|---|---|
| $\alpha$ | $\alpha$ | $\tau_0$ (GPa) | $c$ | $\sigma_c$ (GPa) |
| 2.044 | -1.022 | 0.025 | 7.414 | -0.157 |

JH-1 model combined the linear segment description of the strength and allowed the strength to drop suddenly as the damage evolves. The JH-B model describes the damaged material in a similar way as JH-1 except that an analytical form was used to describe the strength of the intact and failed material and permit the phase change in the damaging process. The JH-2 model incorporated a dimensionless analytical description of the strength and permitted the strength to degrade gradually as the damage was accumulated. Our experimental results[16] indicate that the effect of damage to the load-bearing capacity of the material is abrupt in nature, which is closer to the damage description in JH-1. The JH-1 model is then employed to describe the response of the damaged SiC-N. Most material parameters of SiC can be obtained from Holmquist.[14,15] Figure 7 integrates the strength versus pressure for test data for the intact SiC and JH-1 Model with current data for the failed SiC-N. The slope of the failed SiC-N was determined and listed in Table 1. One data point in Fig. 7 from the intact SiC-N with very small confinement is consistent with JH-1 model description. Due to their smaller values, the data points under 26 MPa and 56 MPa confinement stresses were overlapped together in Fig. 7. Note that the data summarized in Fig. 7 are collected from different SiC's.

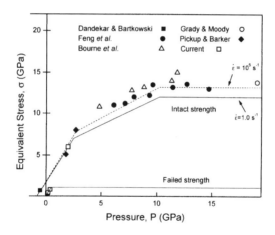

Figure 7. Strength versus pressure for SiC from test data and JH-1 Model.

DISCUSSIONS

Mohr-Coulomb criterion was employed to describe the damaged ceramic under dynamic tri-axial compression. However, it should be noted that the dynamic damage evolution is complicated and the damage is clearly not homogeneous. Mohr-Coulomb model and its equivalent form predict that the axial stress $\sigma_A$ has a linear relationship with the transversal stress $\sigma_T$, while other variables such as damage and strain rate are embedded in material constants. For the failed SiC-N, the equivalent stress in JH-1 model also has a linear relationship with the hydrostatic pressure. The JH-1 model is equivalent to Mohr-Coulomb's description for this particular case.

The damage description is a critical link in the damage modeling of the material. Although indirectly obtained from the crack pattern of post-tested specimens, the damage level should be described in a suitable form such as the micro-crack density, energy or changes in elastic constants, which depends on the corresponding damage evolution theory[36]. However, it is a challenge to cast the damage evolution observed in this study into any of these forms. As for the critical damage level, which also can be qualitatively determined from the crack pattern description, an analytical description that is consistent with the experimental observation is still desired. The damage level determined from the JH model may be a suitable approach, which is defined in term of analytical plastic strain, and not based on micro-mechanical models. The damage levels of the failed SiC-N were uniformly taken as 1.0 in this paper.

In the double-pulse SHPB experiments, only the strength slope of failed ceramic was obtained while the cap strength of failed ceramic was not achieved. The reason is that the hydrostatic pressure and stress is typically small due to low confinements available from the current experimental techniques. As a passive confinement, thin sleeves were employed to confine the ceramic. If a thick sleeve is used, the confining pressure is difficult to evaluate and the effect of the sleeve deduction contribution is not easily achieved. Furthermore, the highly confined ceramic specimen may not crack at all under the axial impact loading because the platens will fail first. Active confinement method such as hydraulic pressure up to 1~2 GPa may be a significant improvement in experimental technique development. However, how to fail the confined specimen remains a challenge.

SUMMARY AND CONCLUSIONS

The crack patterns observed in axially compressed SiC-N specimens show that axial splitting is the main failure mechanism during the failure of the specimen. The slightly damaged SiC-N specimens display only axial cracks at low density. When the specimen is damaged to a nearly critical level, the damage pattern becomes a combination of axial splitting and localized comminution. When the specimen is heavily damaged, the comminution zone increases significantly around broken particles that are further cracked. SEM observations show that the microstructural fracture is intergranular on the surfaces created by axial splitting. Mohr-Coulomb criterion was successfully employed to describe the strength of the damaged ceramic under dynamic tri-axial compression. The parameters of models were determined for damaged and failed SiC-N. The strength slope of the failed SiC-N in JH-1 model was also determined with three stress-pressure data points. Due to the linearity in stress-pressure relation, the JH-1 model description is equivalent to Mohr-Coulomb for the failed SiC-N.

REFERENCES

[1]W. A. Gooch,, "An Overview of Ceramic Armor Application"; pp.3-21 in *Ceramic Transaction*, Vol.**134**, *Ceramic Armor Materials by design*, Edited by C. W. McCauley, A. Crowson, and W.A. Grooch Jr., *et al*. American Ceramic Society, Westerville, OH, 2001.

[2]B. James and C. Lane, "Practical Issues in Ceramic Armor Design"; pp.33-44 in *Ceramic Transaction*, Vol.**134**, *Ceramic Armor Materials by design*, Edited by C. W. McCauley, A. Crowson, and W.A. Grooch Jr., *et al*. American Ceramic Society, Westerville, OH, 2001.

[3]D. A. Shockey, A. H. Marchand, S. R., Skaggs, "Failure Phenomenology of Confined Ceramic Targets and Impacting Rods", *Inter. J. Impact Eng.*, **9**[3], 263-273 (1990).

[4]C. J. Shih, V. F. Nesterenko and M. A. Meyers, "High-strain-rate Deformation and Comminution of Silicon Carbide", *J. Appl. Phys.*, **83**[9], 4660-71 (1998).

[5]J. C. LaSalvia, "Recent Progress on the Influence of Microstructure and Mechanical Properties on Ballistic Performance", pp557-70 in *Ceramic Transaction*, Vol.**134**, *Ceramic Armor Materials by design*, Edited by C. W. McCauley, A. Crowson, and W.A. Grooch Jr., *et al.* American Ceramic Society, Westerville, OH, 2001.

[6]W. Chen and G. Ravichandran, "Static and Dynamic Compressive Behavior of Aluminum Nitride under Moderate Confinement," *J. Am. Ceram. Soc.*, **79**[3], 579-84 (1996).

[7]R. W. Klopp and D. A. Shockey, "The Strength Behavior of Granulated Silicon Carbide at High Strain Rate and Confining Pressure," *J. Appl. Phys.*, **70**[12], 7318-26 (1991).

[8]P. Lundberg and B. Lundberg, "Transition between Interface Defeat and Penetration for Tungsten Projectiles and Four Silicon Carbide Materials," *Int. J. Impact Eng.*, **31**[7], 781-92 (2005).

[9]C. J. Shih and M. A. Meyers, "Damage Evolution in Dynamic Deformation of Silicon Carbide", *Acta Mater.*, **48**[9], 2399-2420 (2000).

[10]T. Jiao, Y. Li and K. T. Ramesh, "High Rate Response and Dynamic Failure of Structural Ceramics," *Int. J. Appl. Ceram. Tech.*, **1**[3], 243-53 (2004).

[11]H. Wang and K. T. Ramesh, "Dynamic Strength and Fragmentation of Hot-pressed Silicon Carbide under Uniaxial Compression," *Acta Mater.*, **52**[2], 355-67 (2004).

[12]W. Chen and H. Luo, "Dynamic Compressive Responses of Intact and Damaged Ceramics from a Single Split Hopkinson Pressure Bar Experiment," *Exp. Mech.*, **44**[2], 295-99 (2004).

[13]H. Luo and W. Chen, "Dynamic Compressive Response of Intact and Damaged AD995 Alumina," *Int. J. Appl. Ceram. Tech.*, **1**[3], 254-60 (2004).

[14]T. J. Holmquist and G. R. Johnson, "Characterization and Evaluation of Silicon Carbide for High-velocity Impact," *J. Appl. Phys.*, **97**[09], 093502/1-12 (2005).

[15]T. J. Holmquist and G. R. Johson, "Response of Silicon Carbide to High Velocity Impact," *J. Appl. Phys.*, **91**[9], 5858-66 (2002).

[16]H. Luo, W. Chen and A. M. Rajendran, "Dynamic Compressive Response of Damaged and Interlock SiC-N Ceramics," *J. Am. Ceram. Soc.*, **89 [1]**, 266-273 (2006).

[17]H. Luo, "Experimental and Analytical Investigation of Dynamic Compressive Behavior of Intact and Damaged Ceramics," Ph.D Dissertation, Univ. of Arizona, Tucson, AZ (2005).

[18]W. Chen and G. Ravichandran, "Failure Mode Transition in Ceramics under Dynamic Multiaxial Compression," *Int. J. Fract.*, **101**[1-2], 141-59 (2000).

[19]S. Satapathy and S. Bless, "Cavity Expansion Resistance of Brittle Materials Obeying a Two-curve Pressure-Shear Behavior", *J. Appl. Phys.*, **88**[7], 4004-12 (2000).

[20]S. Satapathy, "Dynamic Spherical Cavity Expansion in Brittle Ceramics," *Int. J. Solids Struct.*, **38**[32-33], 5833-45 (2001).

[21]C. H. Simha, S. J. Bless and A. Bedford, "Computational Modeling of the Penetration Response of a High-purity Ceramic," *Int. J. Impact Eng.*, **27**[1], 65-86 (2002).

[22]M. F. Ashby and S. D. Hallam, "The Failure of Brittle Solids Containing Small Cracks under Compressive Stress States," *Acta Metal.*, **34**[3], 497-510 (1986).

[23]M. F. Ashby and C. G. Sammis, "The Damage Mechanics of Brittle Solids in Compression," *PAGEOPH*, **133**[3], 489-521 (1990).

[24]H. Horii and S. Nemat-Nasser, "Brittle Failure in Compressive: Splitting, Faulting and Brittle-ductile Transition", *Phil. Trans. Royal Soc.*, **A319**, 337-74 (1986).

[25]A. M. Rajendran and D. J. Grove, "Modeling the Shock Response of Silicon Carbide, Boron carbide, and Titanium Diboride," *Int. J. Impact Eng.*, **18**[6], 611-31(1996).

[26]A. M. Rajendran, "Historical Perspective on Ceramic Materials Damage Models," pp181-89 in *Ceramic Transaction*, Vol.**134**, *Ceramic Armor Materials by design*, Edited by C. W. McCauley, A. Crowson, and W.A. Grooch Jr., *et al*. American Ceramic Society, Westerville, OH, 2001.

[27]K. Bhattacharya, M. Ortiz and G. Ravichandran, "Energy-based Model of Compressive Splitting in Heterogeneous Brittle Solids," *J. Mech. Phys. Solids*, **46**[10], 2171-2181 (1998).

[28]C. Huang and G. Subhash, "Influence of Lateral Confinement on Dynamic Damage Evolution during Uniaxial Compressive Response of Brittle Solids," *J. Mech. Phys. Solids*, **51**[6], 1089-105 (2003).

[29]W. J. Drugan, "Dynamic Fragmentation of Brittle Materials: Analytical Mechanics-based Models," *J. Mech. Phys. Solids*, **49**[6], 1181-208 (2001).

[30]F. Hild, C. Denoual, P. Forquin and X. Brajer, "On the Probabilistic-deterministic Transition Involved in a Fragmentation Process of Brittle Materials," *Comput. Struct.*, **81**[12], 1241-53 (2003).

[31]C. Denoual and F. Hild, "A Damage Model for the Dynamic Fragmentation of Brittle Solids," *Comput. Methods Appl. Mech. Eng.*, **183**[3-4], 247-58 (2000).

[32]D. Templeton, "A Comparison of Ceramic Material Models," pp. 299-308 in *Ceramic Transaction*, Vol.**134**, *Ceramic Armor Materials by design*, Edited by C. W. McCauley, A. Crowson, and W.A. Grooch Jr., *et al*. American Ceramic Society, Westerville, OH, 2001.

[33]D. J. Grove and A. M. Rajendran, "Overview of the Rajendran-Grove Ceramic Failure Model," pp. 371-82 in *Ceramic Transaction*, Vol.**134**, *Ceramic Armor Materials by design*, Edited by C. W. McCauley, A. Crowson, and W.A. Grooch Jr., *et al*. American Ceramic Society, Westerville, OH, 2001.

[34]T. J. Holmquist, D. W. Templeton and K. D. Bishnoi, "Constitutive Modeling of Aluminum Nitride for Large Strain, High-strain Rate, and High-pressure Applications," *Int. J. Impact Eng.*, **25**[3], 211-31 (2001).

[35]G. R. Johnson and T. J. Holmquist (1999), "Response of Boron Carbide Subjected to Large Strains, High Strain Rates, and High Pressures," *J. Appl. Phys.*, **85**[12], 8060-73 (1999).

[36]G. R. Johnson, T. J. Holmquist and S. R. Beissel, "Response of Aluminum Nitride (including a Phase Change) to Large Strains, High Strain Rates, and High pressures," *J. Appl. Phys.*, **94**[3], 1639-46 (2003).

[37]D. Krajcinovic and S. Mastilovic, "Some Fundamental Issues of Damage Mechanics," *Mech. Mater.*, **21**[3], 217-30 (1995).

# GRAIN BOUNDARY CHEMISTRY OF SiC-BASED ARMOR

Edgardo Pabit, Kerry Siebein, and Darryl P. Butt
University of Florida
Materials Science and Engineering Department
Gainesville, Florida 32611

Helge Heinrich
University of Central Florida
AMPAC
Orlando, Florida 32816

Darin Ray, Sarbjit Kaur, R. Marc Flinders, and Raymond A. Cutler
Ceramatec, Inc.
2425 South 900 West
Salt Lake City, Utah 84119

ABSTRACT
    Fourteen SiC-based materials were processed by hot pressing. The single-edge precracked beam (SEPB) fracture toughness varied from 2.4 to 6.8 MPa-m$^{1/2}$ and fracture modes changed from transgranular to intergranular. Grain boundaries and triple points were analyzed using high-resolution transmission electron microscopy combined with energy dispersive spectroscopy (EDS) and electron energy loss spectroscopy (EELS) using an energy-filtered approach. The objective of this study was to compare the grain boundary chemistry of these materials and determine how it affected fracture mode.
    In all samples containing Al, AlN or $Al_2O_3$, oxygen was associated with aluminum at the grain boundaries. Chemical analysis of the hot pressed bulk material showed that O levels ranged between 0.3 and 1.1 wt. %, whereas N levels were as high as 1.5 %. In all of the aluminum-containing samples, the Al and O concentrated at triple points and penetrated along most grain boundaries regardless of the fracture mode. Nitrogen in the AlN-doped samples was difficult to detect by EDS or EELS at 2200°C, suggesting that it diffused into the 4H or 6H polytypes during sintering. Likewise, it was generally difficult to detect the O or Al in solid solution within the SiC grains.
    Simultaneous boron and carbon additions lowered the fracture toughness when either Al or AlN were used at the same cation content. Triple point and grain boundary chemistries, however, were similar to those where no B and C were added. This was primarily due to the enhanced early densification, which resulted in higher amounts of O and N in these compositions. Quantitative fracture mode did not correlate with grain size and grain boundary chemistry, as had been expected.

## INTRODUCTION

    Silicon carbide became a material of interest for armor applications when Svante Prochazka demonstrated the ability to pressureless sinter SiC with B and C additions.[1] This material is useful in applications requiring corrosion resistance, such as seals and bearings, due to clean grain boundaries.[2] As shown by the early work of Alliegro, et al.[3] SiC can be densified with small additions of Al when pressure is applied. Suzuki et al.[4] added $Al_2O_3$ in order to pressureless sinter SiC with enhanced strength. Ezis[5] added AlN to α-SiC in order to make an equiaxed microstructure with a high degree of intergranular fracture. Recent work by Flinders, et al.[6] demonstrated the ability to tailor mechanical properties in SiC-based ceramics containing Al, AlN, or $Al_2O_3$. A rich interplay of microstructure and mechanical properties is possible by controlling chemistry. Pabit, et al.[7] showed by transmission electron microscopy (TEM) and EDS that samples sintered with Al or AlN additives contained O and N independent of whether $B_4C$ and C were added. The purpose of the present work was to look at a wide variety of compositions made with primarily Al or AlN additives in order to understand how grain boundary chemistry affects room-temperature quasi-static mechanical properties.

EXPERIMENTAL PROCEDURES
Processing and Characterization at Ceramatec
The starting powders used for processing were α-SiC (H. C. Stark grade UF-15), Al (Valimet grade H-3), AlN (Tokuyama Soda grade F), $Al_2O_3$ (Sasol North America grade SPA-0.5), $B_4C$ (H. C. Starck grade HS), and C (Capital Resin grade CRC-720) with properties as disclosed previously.[6] All compositions were prepared by batching 600 grams of powder in two-liter high-density polyethylene (HDPE) jars filled with 1.6 kg solid state sintered SiC media and 700 g reagent grade acetone. The slurries were mixed for 16 hours on a ball mill in order to disperse the agglomerates and the powders were stir dried before screening through an 80-mesh screen. The SiC batched with $B_4C$ and C was pyrolyzed by heating in $N_2$ to 600°C. Billets (45 mm x 45 mm x 6 mm) were hot pressed at 28 MPa inside graphite dies by heating to 1500°C in a vacuum of 1 torr, holding at 1500°C for one hour to remove SiO and CO from the samples, and then backfilling with Ar and heating to temperatures between 1900 and 2200°C for one hour. Table I lists the compositions and their respective processing conditions.

**Table I**
**Materials Evaluated**

| Designation | Composition (wt. %) | Hot Pressing Conditions |
|---|---|---|
| 1.65Al-1900°C | α-SiC-1.65Al | 1900°C-1hr |
| 1.65Al-2100°C | α-SiC-1.65Al | 2100°C-1hr |
| 1.65Al-2200°C | α-SiC-1.65Al | 2200°C-1hr |
| 1.65Al-0.5B₄C-2C-2100°C | α-SiC-1.65Al-0.5B₄C-2C | 2100°C-1hr |
| 3.3Al-2000°C | α-SiC-3.3Al | 2000°C-1hr |
| 3.3Al-2200°C | α-SiC-3.3Al | 2200°C-1hr |
| 2.5AlN-2100°C | α-SiC-2.5AlN | 2100°C-1hr |
| 2.5AlN-2200°C | α-SiC-2.5AlN | 2200°C-1hr |
| 2.5AlN-0.5B₄C-2100°C | α-SiC-2.5AlN-0.5B₄C | 2100°C-1hr |
| 2.5AlN-0.5B₄C-2C-2100°C | α-SiC-2.5AlN-0.5B₄C-2C | 2100°C-1hr |
| 5AlN-2000°C | α-SiC-5AlN | 2000°C-1hr |
| 5AlN-2200°C | α-SiC-5AlN | 2200°C-1hr |
| 3.1Al₂O₃-2100°C | α-SiC-3.1Al₂O₃ | 2100°C-1hr |
| 0.5 B₄C-2C-2100°C | α-SiC-2C-0.5 B₄C | 2100°C-1hr |

The hot pressed billets were ground with a 320 grit diamond wheel to make 3 mm x 4 mm x 45 mm bars as specified by ASTM C-1421-99.[8] Toughness was measured using the single-edge precracked beam (SEPB) technique using black printer ink to mark cracks as described previously.[9] All crack planes were parallel to the hot pressing direction. Each data point is the mean of five bars tested, with error bars representing one standard deviation.

A microhardness machine (Leco model LM-100) was used to obtain Vickers and Knoop hardness data on polished SEPB bars. Data were taken at a load of 9.8 N. Each data point represents the mean of ten measurements, with error bars representing one standard deviation.

Rietveld analysis[10,11] was used to determine SiC polytypes present in the densified samples with X-ray diffraction patterns collected from 20-80° 2-theta, with a step size of 0.02°/step and a counting time of 4 sec/step.

Polished surfaces were etched with boiling Murakami's reagent,[6] modified Murikami's reagent,[12] $CF_4$-$O_2$ plasma,[13] or thermally etched dependent on composition and grain size. Grain size was determined by the line-intercept method, where the multiplication constant ranged between 1.5 (equiaxed grains) and 2.0 (elongated, plate-shaped grains).[14] Typically, 400-600 grains were measured for each composition in order to get a mean grain size.

The aspect ratios of the three most acicular grains in each of five micrographs were used to estimate a comparative aspect ratio. The fracture mode was determined from polished, precracked SEPB bars. The SEPB bars were subsequently etched to get a quantitative estimate of the fracture mode by viewing the crack path over a distance of 35-250 μm, depending on grain size. Oxygen and nitrogen were analyzed on selected starting powders and sintered samples using ASTM method E1409 for the determination of oxygen by the inert gas fusion/thermal conductivity detection technique with a Leco TC-136 model analyzer.

## TEM Characterization at University of Florida

Samples were ion-beam thinned and imaged with a transmission electron microscope (JEOL model 2010F) operated at 200 kV with a resolution limit of 0.2 nm. Chemical analysis was obtained by EDS using the scanning transmission mode with a probe size of approximately 8 nm. Energy filtered transmission electron microscopy (EFTEM) with a three-window technique[15] for forming electron spectroscopic images by filtering out the background radiation[16] was also used. A field-emission TEM (Technai model F30) at an accelerating voltage of 300 kV operated at the University of Central Florida was used for all EFTEM imaging.

## RESULTS AND DISCUSSION
### Sample Characterization

Table II gives density, polytypes present after processing based on the Rietveld fitting of XRD patterns, grain size and aspect ratio, as well as chemical analysis of the densified samples. All materials without carbon as an additive reached greater than 99.5 % of their theoretical density. No cubic (3C) or hexagonal 2H polytypes were noted by Rietveld analysis. Aluminum is known[17] to promote the 6H to 4H transformation, consistent with the results shown in Table II for the SiC-Al samples. Temperature accelerates the transformation, which increases both the grain size and their aspect ratio. The addition of $B_4C$ and C to SiC-Al retards the conversion to 4H but still promotes grain growth. The main drawback of this approach is the lower densification which results. Based on the work of Zangvil and Ruh[18] one could expect a 4H solid solution to occur between SiC and AlN. The SiC-2.5 wt. % AlN samples have the same Al content as the SiC-1.65 wt. % Al compositions. When comparing identical Al contents and

**Table II**
**Density, Polytypes, Grain Size, Aspect Ratio, and Chemical Analysis**

| Designation | Density (g/cc) | Polytypes Present 4H | 6H | 15R | Grain Size (μm) | Aspect Ratio | Chemical Analysis (wt. %) O | N |
|---|---|---|---|---|---|---|---|---|
| 1.65Al-1900°C | 3.22 | 8.1 | 83.5 | 8.5 | 1.1±0.1 | 2.9±0.5 | 1.01 | 0.01 |
| 1.65Al-2100°C | 3.20 | 61.7 | 33.0 | 5.4 | 2.2±0.3 | 3.2±0.6 | 0.76 | 0.01 |
| 1.65Al-2200°C | 3.21 | 66.7 | 28.5 | 4.8 | 3.1±0.2 | 4.2±1.1 | 0.46 | 0.01 |
| 1.65Al-0.5B₄C-2C-2100°C | 3.16 | 21.2 | 71.7 | 0.0 | 6.2±1.0 | 3.5±1.5 | 0.98 | 0.01 |
| 3.3Al-2000°C | 3.19 | 12.7 | 79.7 | 5.9 | 1.9±0.2 | 2.5±0.6 | 0.80 | 0.01 |
| 3.3Al-2200°C | 3.18 | 76.0 | 24.0 | 0.0 | 5.1±0.2 | 6.5±1.6 | 0.27 | 0.01 |
| 2.5AlN-2100°C | 3.22 | 26.8 | 62.2 | 7.7 | 1.2±0.1 | 2.2±0.2 | 0.44 | 0.20 |
| 2.5AlN-2200°C | 3.21 | 39.7 | 53.0 | 7.2 | 1.9±0.1 | 2.3±0.7 | 0.64 | 0.42 |
| 2.5AlN-0.5B₄C-2100°C | 3.20 | 52.3 | 44.0 | 3.7 | 3.1±0.2 | 2.7±0.7 | 0.38 | 0.84 |
| 2.5AlN-0.5B₄C-2C-2100°C | 3.16 | 49.4 | 45.7 | 4.9 | 2.5±0.3 | 2.3±0.4 | 0.72 | 0.59 |
| 5AlN-2000°C | 3.19 | 0.8 | 89.5 | 9.7 | 0.5±0.1 | 4.1±1.1 | 0.72 | 1.45 |
| 5AlN-2200°C | 3.22 | 6.6 | 82.5 | 10.9 | 1.1±0.1 | 2.5±0.2 | 0.49 | 0.89 |
| 3.1Al₂O₃-2100°C | 3.21 | 9.7 | 83.1 | 7.2 | 1.0±0.1 | 2.7±0.6 | 1.08 | 0.00 |
| 0.5B₄C-2C-2100°C | 3.19 | 0.0 | 94.8 | 5.2 | 2.4±0.2 | 3.2±0.9 | 0.02 | 0.01 |

temperatures, it is clear from Table II, that AlN retards the formation of 4H polytypes compared to Al. It is well known that nitrogen limits grain growth, which is also evident from the Table II data. Similar to SiC-Al, when B₄C and C are added to SiC-AlN, the density decreases. This is

**Table III**
**Mechanical Characterization**

| Designation | % Intergranular Fracture | Toughness (MPa√m) | Hardness (GPa) HK1 | HV1 |
|---|---|---|---|---|
| 1.65Al-1900°C | 50 | 4.0±0.2 | 20.1±0.2 | 25.3±0.7 |
| 1.65Al-2100°C | 52 | 4.7±0.4 | 19.2±0.2 | 22.1±0.7 |
| 1.65Al-2200°C | 39 | 5.7±0.1 | 18.9±0.2 | 20.8±0.3 |
| 1.65Al-0.5B₄C-2C -2100°C | 30 | 3.1±0.1 | 18.2±0.5 | 20.3±0.3 |
| 3.3Al-2000°C | 46 | 4.2±0.1 | 18.9±0.2 | 22.1±0.6 |
| 3.3Al-2200°C | 57 | 6.8±0.1 | 18.5±0.1 | 20.5±0.5 |
| 2.5AlN-2100°C | 88 | 3.5±0.2 | 19.8±0.3 | 21.9±0.3 |
| 2.5AlN-2200°C | 38 | 3.5±0.4 | 19.7±0.2 | 22.9±0.1 |
| 2.5AlN-0.5B₄C-2100°C | 20 | 3.4±0.4 | 18.7±0.2 | 21.0±0.6 |
| 2.5AlN-0.5B₄C-2C-2100°C | 25 | 2.8±0.2 | 19.0±0.2 | 22.7±0.3 |
| 5AlN-2000°C | 84 | 3.9±0.2 | 21.2±0.2 | 25.0±0.4 |
| 5AlN-2200°C | 87 | 4.2±0.1 | 20.9±0.2 | 24.1±0.6 |
| 3.1Al₂O₃-2100°C | 84 | 3.8±0.2 | 20.7±0.1 | 23.5±0.7 |
| 0.5B₄C-2C-2100°C | 29 | 2.4±0.1 | 20.8±0.1 | 25.7±0.4 |

readily apparent when one compares the density with the SiC sintered with just $B_4C$ and C, where carbon aids in removal of silica. Since Al or AlN is also capable of reacting with $SiO_2$, the C is not necessary for densification.

Oxygen and nitrogen analysis in the densified materials is also shown in Table II. The starting lot of SiC was analyzed to have 1.2 wt. % oxygen. The Al starting powder was measured at 0.76 wt. % O and the AlN at 1.3 wt. % O. The $Al_2O_3$ powder contained 0.11 wt. % N, in addition to its theoretical 31.4 wt. % O. As shown in Table II, the phenolic resin in the SiC-0.5 wt. % $B_4C$-2 wt. %C material was effective in removing the oxygen, but whenever an Al compound was present there was significant oxygen present. Al was more effective than $Al_2O_3$ in retaining oxygen. As expected, the amount of retained O decreased with increasing hot pressing temperature. The amount of retained nitrogen in SiC-AlN samples was different from that of oxygen in SiC-Al compositions. Increasing Al additions slow down diffusion and result in less retained oxygen. However, increasing the amount of AlN, which limits grain growth and allows for closed porosity to be attained sooner, allows nitrogen to be retained. Boron carbide, with or without C, also results in high amounts of retained nitrogen when AlN is the sintering additive since it also allows closed porosity to be attained at lower temperatures. Simply retaining oxygen or nitrogen in the sample, however, does not guarantee that the material fractures intergranularly as seen by comparing the data in Tables II and III. For example, the SiC-1.65 wt. % Al-0.5 wt. % $B_4C$-2 wt. % C sample retained 90 % of its oxygen after densification but fractured primarily transgranularly and had low toughness.

Figure 1 shows the etched microstructures at the same magnification. The elongation in the SiC grains when using Al as a densification aid at high temperatures is evident. This suggests that crack bridging is a likely mechanism for the toughening for these compositions, which is evident from the data in Table III. The grains in the SiC-Al densified at 1900-2000°C, as well as the other compositions, appear to be fairly equiaxed in shape.

Figures 2 and 3 give different views of the fracture mode. Figure 2 shows fracture surfaces at constant magnification, whereas Figure 3 compares Vicker's indents where the fracture mode can be inferred from the crack pattern. These are both different from the quantitative data in Table III which were assessed by monitoring the surface crack pattern behind the crack tip after etching a polished, precracked surface. By comparing all three methods of assessing fracture mode, it is clear that the SiC-0.5 wt. % $B_4C$-2 wt. %C material is the most transgranular in fracture mode, consistent with its low fracture toughness. Using the same method of evaluation, the SiC-2.5 wt. % AlN at 2100°C was the most intergranular. It did not have the highest toughness, however, since the mixed-mode SiC-Al samples with elongated grains were better able to bridge the cracks.

The data in Tables II and III were used to fit existing models for toughening. While it is clear that crack bridging leads to toughening in SiC, the data in Tables II and III have some limitations. For example, the aspect ratios are only those for the most elongated grains and are not representative of the majority of the grains. The % intergranular fracture was measured away from the crack tip, where the bridging is most likely to occur. Despite these limitations, the approach of Zenotchkine, et al.[19] of correlating fracture toughness with the product of the aspect ratio times the square root of grain size times the volume fraction of the bridging grains, as shown in Figure 4, was used to compare all of the compositions assuming that all of the intergranular grains bridge the crack. The two data points which prevent a modest linear fit are noted on the graph. Both of these compositions have little intergranular fracture. Investigation of the grain boundary phases in these materials is of interest.

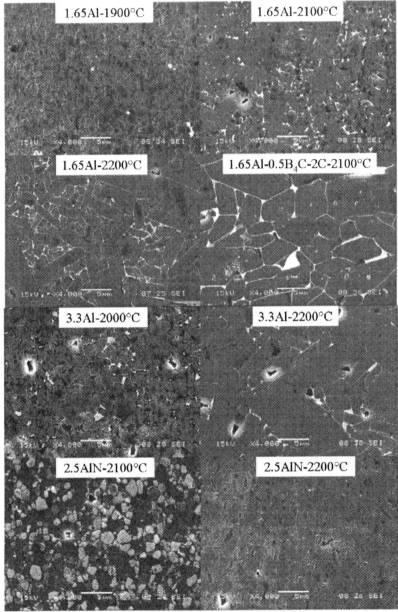

Figure 1. Etched microstructures of compositions listed in Table I. All markers are 5 µm.

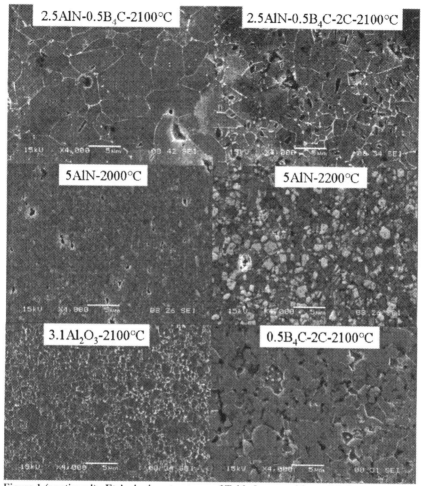

Figure 1 (continued). Etched microstructures of Table I compositions. Markers are 5 μm.

Transmission Electron Microscopy

High-resolution TEM was used to assess grain boundaries, which were 1-2 nm in width, making it difficult to use EDS to determine the chemistry of the grain boundaries. The EFTEM which uses EELS was much more effective for showing what was at the grain boundaries. Figure 5 shows an EFTEM image where it is clear that Al and O are enriched along the grain boundary of SiC-1.65 wt. % Al sample hot pressed at 2100°C. It is more obvious from EFTEM that aluminum and oxygen segregate in the liquid whereas nitrogen is well distributed in the bulk (see Figures 6 and 7). However, at high concentrations of AlN, it is obvious that nitrogen is also

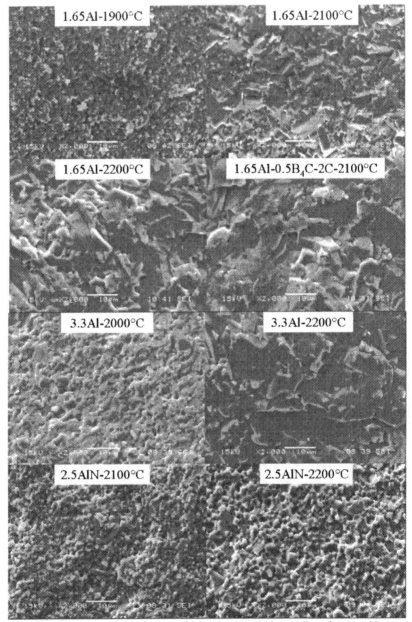

Figure 2. Fracture surfaces of compositions listed in Table 1. All markers are 10 μm.

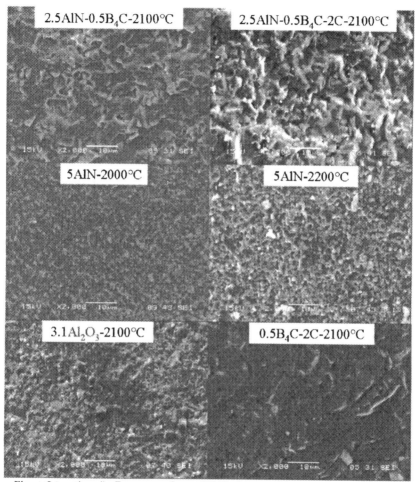

Figure 2 (continued). Fracture surfaces of Table I compositions. Markers are 10 μm.

present in the liquid phase (see Figure 8). When the same composition is processed at 2200°C the nitrogen is no longer observed by EFTEM, consistent with the higher temperature allowing diffusion of the nitrogen into the bulk. For SiC-N, a commercial material hot pressed for long times below 2000°C there is no obvious concentration of N at grain boundaries or triple points, although Al and O are concentrated at triple junctions[20] consistent with nitrogen solubility at lower temperatures.[18] Surprisingly, the SiC-2.5 wt. % AlN-0.5 wt. % B₄C-2 wt. %C composition with low toughness and primarily transgranular fracture showed a triple junction

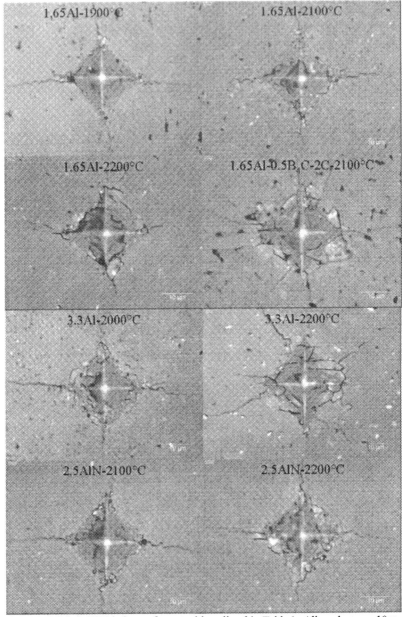

Figure 3. Vickers HV1 indents of compositions listed in Table I. All markers are 10 μm.

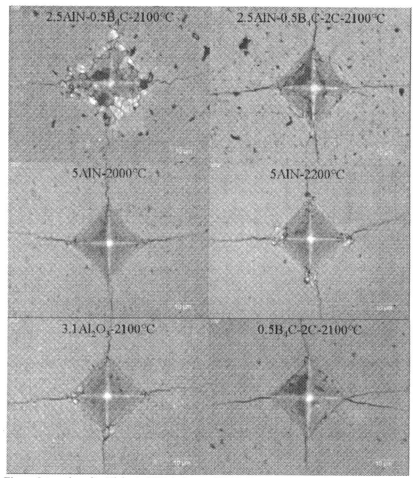

Figure 3 (continued). Vicker's HV1 indents of Table I compositions. Markers are 10 μm.

with a low dihedral angle and Al clearly present in the grain boundaries. Both this composition and the companion SiC-2.5 wt. % AlN-0.5 wt. % $B_4C$ material had B segregated to triple junctions along with Al and O. This is in stark contrast to the SiC-0.5 wt. % $B_4C$-2 wt. % C composition which showed no triple junctions and clean grain boundaries, as expected. From these observations, it is clear that Al and AlN getter the surface silica, reacting to form $Al_2O_3$-rich liquids which wet these materials. It is unclear, however, why fine-grained materials containing these liquids can result in a high degree of transgranular fracture and low fracture

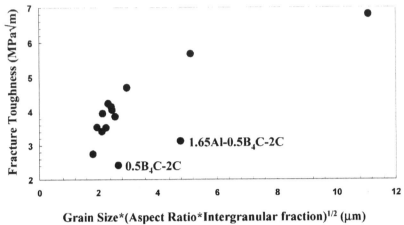

Grain Size*(Aspect Ratio*Intergranular fraction)$^{1/2}$ (μm)

Figure 4. Fracture toughness as a function of the product of grain size times the square root of aspect ratio times the intergranular fraction. A linear fit is expected based on crack bridging.[19]

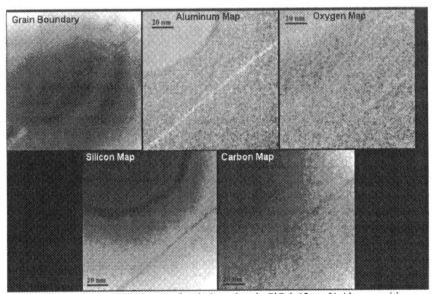

Figure 5. Energy filtered TEM image of grain boundary in SiC-1.65 wt. % Al composition densified at 2100°C. Al and O are enriched at grain boundary while Si and C are depleted.

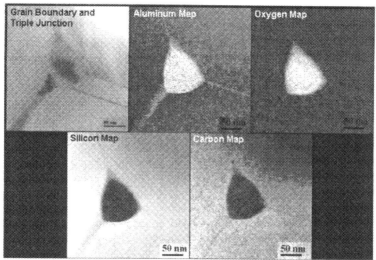

Figure 6. EFTEM of SiC-3.3 wt. % Al hot pressed at 2000°C. Al and O are concentrated at triple points, with Al clearly evident along grain boundaries.

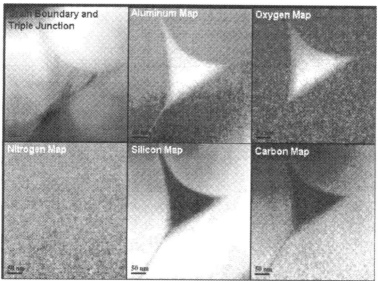

Figure 7. EFTEM of SiC-2.5 wt.% AlN hot pressed at 2200°C. Al and O are concentrated at triple point, with Al clearly evident along grain boundaries. N is not segregated at triple point.

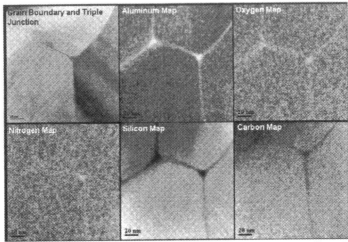

Figure 8. SiC-5.0 wt. % AlN hot pressed at 2000°C. Note that nitrogen, as well as Al and O are present along grain boundaries and triple points.

toughness while companion samples show a high degree of intergranular fracture with similar grain boundary and triple junction chemistries.[20] Further TEM work is necessary to understand the subtle differences in grain boundary chemistries which allow this to occur. The EFTEM was more effective in observing segregation of secondary phases at grain boundaries than EDS, due to the small width of the grain boundaries. Al was readily apparent along most grain boundaries in all of the materials containing Al, AlN, or $Al_2O_3$ as a densification additive. It was not possible using these techniques to observe O or Al in solid solution in the SiC matrix.

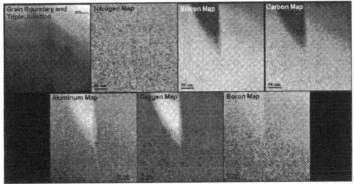

Figure 9. SiC-2.5 wt. % AlN-0.5 wt. % $B_4C$-2 wt. % C densified at 2100°C clearly showing Al penetration from the triple point into the grain boundary. This material showed low toughness (2.8 MPa-m$^{1/2}$) with primarily transgranular fracture despite having a grain boundary phase and relatively small grain size.

CONCLUSIONS

Grain boundary chemistries were surprisingly similar for all materials investigated with the exception of SiC densified with only $B_4C$ and C. Al, AlN, and $Al_2O_3$ resulted in triple junctions with Al and O forming a liquid phase, which in most cases had dihedral angles less than 72.5° allowing penetration of the liquid to occur along grain boundaries. Nitrogen was present in the liquid phase at 2000°C but was generally gone by 2100°C and was not observed at 2200°C. When boron was added in combination with Al or AlN, it was observed in the triple junctions.

Characterization of the microstructure of these materials in correlation with single-edged precracked beam fracture toughness measurements allowed a general understanding of what allows high toughness to be obtained in these materials. In order to obtain a toughness in excess of 5 MPa-m$^{1/2}$ large bridging grains with a moderate to high degree of intergranular fracture are required. Al additions allow this to occur more readily than AlN, using comparable processing, due to the enhanced grain growth and elongated grains which occur when nitrogen is absent. Surprisingly, transmission electron microscopy of grain boundary chemistry did not explain why low toughness occurred in materials which contained Al, O and N in cases where AlN was added. It is likely that subtle differences in wetting characteristics change the nature of the penetration along grain boundaries. More detailed TEM work is necessary in order to discover these differences.

ACKNOWLEDGEMENT

This SBIR work was performed for the U.S. Army under contract DAAD17-02-C-0052. Appreciation is expressed to Lyle Miller of Ceramatec for help with x-ray diffraction work. Technical discussions with Professor Dinesh K. Shetty at the University of Utah are gratefully acknowledged.

REFERENCES

[1]S. Prochazka, "Role of Boron and Carbon in the Sintering of Silicon Carbide," *Proc. of the Conference on Ceramics for High Performance Applications*, Hyannis, Mass., 1973, ed. by J. J. Burke, A. E. Gorum, and R. M. Katz (Brook Hill Publ. Co., 1975).

[2]R. F. Davis, J. E. Lane, C. H. Carter, Jr., J. Bentley, W. H. Wadlin, D. P. Griffis, R. W. Linton, and K. L. More, "Microanalytical and Microstructural Analysis of Boron and Aluminum Regions in Sintered Alpha Silicon Carbide," *Scanning Electron Microscopy*, 3 1161-7 (1983).

[3]R. A. Alliegro, L. B. Coffin, and J. R. Tinklepaugh, "Pressure-Sintered Silicon Carbide," *J. Am. Ceram. Soc.*, 39[11], 386-89 (1956).

[4]K. Suzuki, T. Ono and N. Shinohara, "Dense Sintered Silicon Carbide Ceramic," U. S. Patent 4,354,991 (October 19, 1982).

[5]A. Ezis, "Monolithic, Fully Dense Silicon Carbide Material, Method of Manufacturing, and End Uses," U. S. Patent 5,372,978 (Dec. 13, 1994).

[6]R. M. Flinders, D. Ray, M. A. Anderson, and R. A. Cutler, "Microstructural Engineering of the Si-C-Al-O-N System," *Ceram. Trans.*, 178, 63-78 (2005).

[7]E. Pabit, S. Crane, K. Siebein, D. P. Butt, D. Ray, M. Flinders, and R. A. Cutler, "Grain Boundary and Triple Junction Chemistry of Silicon Carbide with Aluminum or Aluminum Nitride Additive," *Ceram. Trans.*, 178, 91-102 (2005).

[8]ASTM C 1421-99, Standard Test Methods for Determination of Fracture Toughness of Advanced Ceramics at Ambient Temperature, pp. 641-672 in *1999 Annual Book of Standards* (ASTM, Philadelphia, PA 1999).

[9]D. Ray, M. Flinders, A. Anderson, and R. A. Cutler, "Hardness/Toughness Relationship for SiC Armor," *Ceram. Sci. and Eng. Proc.*, **24**, 401-10 (2003).

[10]H. M. Rietveld, "A Profile Refinement Method in Neutron and Magnetic Structures," *J. Appl. Crystallogr.*, **2**, 65-71 (1969).

[11]D. L. Bish and S. A. Howard, "Quantitative Phase Analysis Using the Rietveld Method," *J. Appl. Crystallogr.*, **21**, 86-91 (1988).

[12]D. H. Stutz, S. Prochazka, and J. Lorenz, "Sintering and Microstructure Formation of b-Silicon Carbide," *J. Am. Ceram. Soc.*, **68**[9], 479-82 (1985).

[13]M. Flinders, D. Ray, A. Anderson, and R. A. Cutler, "High-Toughness Silicon Carbide as Armor," *J. Am. Ceram. Soc.*, **88**[8], 2217-2226 (2005).

[14]E. E. Underwood, *Quantitative Stereology*, (Addison-Wesley, Reading, MA. 1970).

[15]F. Hofer, P. Warbichler, and W. Grogger, "Imaging of Nanometer-Sized Precipitates in Solids by Electron Spectroscopic Imaging," *Ultramicroscopy*, **59**, 15-31 (1995).

[16]F. Hofer, W. Grogger, G. Kothleitner, and P. Warbichler, "Quantitative Analysis of EFTEM Elemental Distribution Images," *Ultramicroscopy*, **67**, 83-103 (1997).

[17]R. M. Williams, B. N. Juterbock, S. S. Shinozaki, C. R. Peters, and T. J. Whalen, "Effects of Sintering Temperatures on the Physical and Crystallographic Properties of β-Silicon Carbide," *Am. Ceram. Soc. Bull.*, **64**[10], 1385-9 (1985).

[18]A. Zangvil and R. Ruh, "Phase Relationships in the Silicon Carbide-Aluminum Nitride System," *J. Am. Ceram. Soc.*, **71**[10], 884-90 (1988).

[19]M. Zenotchkine, R. Shuba, and I-W Chen, "Effect of Seeding on the Microstructure and Mechanical Properties of α-SiAlON: III, Comparison of Modifying Cations," *J. Am. Ceram. Soc.*, **86**[7], 1168-75 (2003).

[20]E. Pabit, "Grain Boundary and Triple Junction Chemistry of Silicon Carbide Sintered with Minimum Additives for Armor Applications," Ph.D. Dissertation, University of Florida (2005).

# EFFECT OF MICROSTRUCTURE AND MECHANICAL PROPERTIES ON THE BALLISTIC PERFORMANCE OF SiC-BASED CERAMICS

Darin Ray, R. Marc Flinders, Angela Anderson, and Raymond A. Cutler
Ceramatec, Inc.
2425 South 900 West
Salt Lake City, Utah, 84119

James Campbell and Jane W. Adams
Army Research Laboratory
Aberdeen, Maryland 21005

ABSTRACT

Processing additives and conditions allow a wide variation in microstructures and mechanical properties for SiC-based ceramics prepared by hot pressing. Five experimental materials with a wide variety of microstructures and mechanical properties were fabricated and compared with SiC-N, a commercially available material. Quasi-static fracture modes varied from predominantly transgranular to primarily intergranular, leading to a twofold increase in the single-edged precracked beam (SEPB) fracture toughness. Hardness varied due to an order of magnitude change in grain size, while porosity varied by less than one percent. Ballistic $V_{50}$ performance was measured using 14.5 mm projectiles shot at ceramic/composite targets. The relative ranking of ballistic performance is discussed in terms of microstructure and mechanical properties.

The data indicate that a wide variety of SiC-based materials can give good ballistic results, contradicting some of the theories about what is important to improve ceramics for armor. Several materials were as good or better than SiC-N in these tests. The relative ballistic ranking was not predictable based solely on hardness, toughness, strength, grain size, elastic modulus, or fracture mode. Although Weibull modulus correlated with ballistic performance, this is likely coincidental since strength values extrapolated to a low failure probability based on the respective Weibull distributions could not rank the ballistic results.

## INTRODUCTION

SiC is the ceramic armor of choice for moderate to heavy threats due to the likely amorphitization of $B_4C$ at high pressures. The key to making improved armor is to understand what controls the ballistic performance of the armor system. This paper focuses only on the ceramic portion of the ballistic package realizing that the performance is tied to the system. In previous work[1] using 7.62 mm diameter by 51 mm long WC-Co cored ammunition, it was shown that the depth of penetration of a variety of SiC-based ceramics scaled with the hardness of the ceramic, in agreement with early work by Krell and Straussburger.[2] It was also demonstrated, however, that the $V_{50}$ performance for 14.5 mm diameter WC-Co cored ammunition on four different SiC-materials all ranged between 722 and 750 m/s regardless of their room-temperature hardness and toughness.[1] While high Young's modulus appeared to be beneficial for obtaining a high $V_{50}$, no direct correlation existed between them. Since the $V_{50}$ values were lower than the muzzle velocity for the BS-41 bullets used, due to the thickness of the ceramic tiles, it was suggested that testing thicker tiles might allow greater differentiation between materials.[1]

85

Recent work showed that it was possible to tailor microstructures and mechanical properties in SiC-based ceramics by controlling chemistry and processing.[3,4] It was possible to hot press a variety of SiC compositions with different microstructures and mechanical properties to near theoretical density. Five of these compositions were selected for ballistic testing in comparison to SiC-N, a benchmark armor material fabricated by BAE Advanced Ceramic Division. The purpose of this paper is to discuss the processing, characterization, and ballistic test results in an effort to determine what properties are relevant for making improved SiC armor for advanced threats.

EXPERIMENTAL PROCEDURES

The six materials chosen for this study are listed in Table I. SiC-N (BAE lot 6-0229) was supplied by the Army Research Laboratory (ARL) as square tiles. The tiles were ground by Quality Magnetics (Compton, CA) to make round disks 100 mm in diameter by 19 mm thick in order to have a comparable geometry to the parts prepared at Ceramatec, Inc. Alpha-SiC (Starck grade UF-15), Al (Valimet grade H3), AlN (Tokuyama Soda grade F), $B_4C$ (Starck grade HS), and C (Durez grade 7716 phenolic resin assuming a yield of 50% after pyrolysis) were used to prepare the compositions made at Ceramatec. The powders were milled in 4 L high-density polyethylene bottles using SiC media, with a ball to charge ratio of 2.5:1, in acetone for 16 hours. The powders were dried and sieved through a 44 μm screen before hot pressing at 21 MPa for one hour in stagnant Ar under conditions as described in Table I. The hot pressed disks were ground to identical dimensions as SiC-N by the same vendor. Density and ultrasonic velocities were measured on each disk. One disk of each material was used for destructive characterization testing and six disks were supplied to ARL for $V_{50}$ evaluation.

The hot pressed disks were sliced and then ground with a 320 grit diamond wheel to make 3 mm x 4 mm x 45 mm bars as specified by ASTM C-1421-99. Density was measured by water displacement. Fracture toughness was measured using the single-edge precracked beam (SEPB) technique[5] as described previously[6] except that black printer ink was used to mark the crack location. All crack planes were parallel to the hot pressing direction. Each data point is the mean of 9-10 bars tested, with error bars representing one standard deviation.

A microhardness machine (Leco model LM-100) was used to obtain Vickers and Knoop hardness data on polished SEPB bars. Data were taken at a load of 9.8 N. Each data point represents the mean of ten measurements, with error bars representing the standard deviation.

Rietveld analysis[7,8] was used to determine SiC polytypes present in the densified samples with X-ray diffraction patterns collected from 20-80° 2-theta, with a step size of 0.02°/step and a counting time of 4 sec/step. Oxygen and nitrogen were analyzed on sintered samples using

**Table I**
**Materials Evaluated**

| Designation | Composition (wt. %) | Hot Pressing Temperature (°C) |
|---|---|---|
| SiC-N | Proprietary (BAE) | Proprietary (BAE) |
| B-C | SiC-2 % C-0.5 % $B_4C$ | 2150 |
| Al (1950) | SiC-1.65 % Al | 1950 |
| Al (2200) | SiC-1.65 % Al | 2200 |
| AlN | SiC-5 wt. % AlN | 2100 |
| AlN-$B_4C$ | SiC-2.5 % AlN-0.5 % $B_4C$ | 2200 |

ASTM method E1409 for the determination of oxygen by the inert gas fusion/thermal conductivity detection technique with a Leco TC-136 model analyzer.

Polished samples of SiC densified with Al or AlN-$B_4C$ were plasma-etched by evacuating and back-filling with 400 millitorr of $CF_4$-10% $O_2$ and etching for 2-10 minutes. SiC-N and AlN samples were thermally etched at 1540°C for one hour in flowing Ar. The B-C sample was etched in a modified Murakami solution[9] at 75°C for 15 minutes. Grain size was determined by the line-intercept method, where the multiplication constant ranged between 1.5 (equiaxed grains) and 2.0 (elongated, plate-shaped grains).[10] Typically, 400-600 grains were measured for each composition in order to get a mean grain size. The aspect ratios of the three most acicular grains in each of 5 micrographs were used to estimate a comparative aspect ratio.

The fracture mode was determined from polished, precracked SEPB bars. The precracked bars were subsequently etched as described above to get a quantitative estimate of the fracture mode by viewing the crack path over a distance of 50-200 μm, depending on grain size.

Flexural strength was measured on 25 bars (3 mm x 4 mm x 45 mm) using a 40 mm support span and a 20 mm loading span, with the crosshead speed at 0.5 mm/min. A two-parameter Weibull analysis was used to calculate the characteristic strength. Young's modulus was measured in flexure using strain gages.

Ballistic testing was performed at ARL using steel to surround the targets and composite backing and cover plates. The $V_{50}$, in theory, is the velocity of the bullet at which the probability of the projectile penetrating through the aluminum backing plate is 50%. Due to the limited number of targets tested, this value was taken as the mean of the two highest velocity tests at which the bullet did not fully penetrate the composite backing and the two lowest velocity tests at which the bullet fully penetrated the backing.

RESULTS AND DISCUSSION

Materials Characterization

Table II gives density, SiC polytypes, mean grain size, aspect ratio, Young's modulus, as well as oxygen and nitrogen analysis for the six materials described in Table I. Five of the six materials had similar densities, with the SiC densified without an aluminum additive being slightly lower in density at 3.19 g/cc. The exact theoretical densities are difficult to calculate, but polished cross-sections indicate that all materials are greater than 99 % of theoretical. The starting SiC contained 1.0 % oxygen[3], which would require 0.4 wt. % C to remove the surface

**Table II**
**Characterization of Materials for Ballistic Testing**

| Designation | Density (g/cc) | Polytypes Present 4H | 6H | 15R | Grain Size (μm) | Aspect Ratio | E (GPa) | Chemical Analysis (wt. %) O | N |
|---|---|---|---|---|---|---|---|---|---|
| SiC-N | 3.22±0.01 | 0.0 | 100.0 | 0.0 | 2.7±0.2 | 2.7±0.5 | 442±1 | 0.59 | 0.20 |
| B-C | 3.19±0.01 | 0.0 | 91.9 | 8.1 | 3.9±0.4 | 2.9±0.4 | 421±3 | 0.17 | 0.03 |
| Al (1950°C) | 3.21±0.01 | 10.0 | 81.6 | 8.4 | 1.1±0.1 | 2.7±0.3 | 431±1 | 0.85 | 0.01 |
| Al (2200°C) | 3.21±0.01 | 46.5 | 47.5 | 6.0 | 2.7±0.3 | 3.1±0.5 | 428±3 | 0.71 | 0.01 |
| AlN | 3.22±0.01 | 7.6 | 86.0 | 6.4 | 0.8±0.1 | 2.5±0.4 | 429±3 | 0.95 | 1.57 |
| AlN-$B_4C$ | 3.21±0.01 | 59.4 | 40.6 | 0.0 | 3.6±0.2 | 2.8±0.5 | 423±4 | 0.98 | 0.74 |

$SiO_2$ from the SiC as CO and SiO. Al and AlN starting powders contained 0.76 and 1.29 wt. % O, respectively, as impurities[3] resulting in all materials processed at Ceramatec having about one wt. % O prior to heating. Only the SiC densified without Al or AlN had low oxygen after heating. If all of the O within the samples was present as $Al_2O_3$ after densification, then theoretical densities would be in the 3.21-3.23 g/cc range.

Aluminum is expected to react with surface silica on the silicon carbide to form alumina and silicon. In a reducing environment, SiC and $Al_2O_3$ react to form $Al_2O$, CO, and SiO.[11] Al, B, N, and O can go into solid solution with SiC. AlN and $Al_2OC$ are isostructural with hexagonal SiC[12] and one would expect that the starting 6H SiC would form 4H if AlN were in solid solution at 2100-2200°C.[13] However, the composition containing 5 wt. % AlN shows little 4H formation, indicating that a complete solid solution is not formed. Aluminum is known to promote the 6H to 4H transformation,[14] consistent with the results shown in Table II for the SiC-Al sample hot pressed at 2200°C. The addition of 0.5 wt. % $B_4C$ to SiC containing 2.5 wt. %

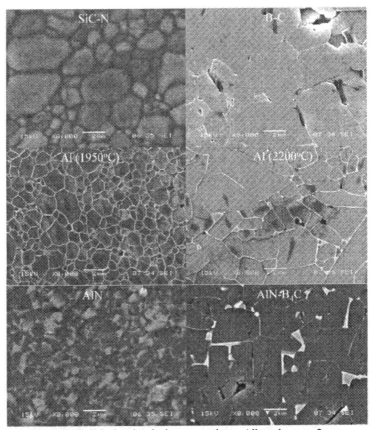

Figure 1. Polished and etched cross-sections. All markers are 2 μm.

AlN was most effective in promoting 4H formation as shown in Table II. No cubic (3C) or 2H polytypes were observed in any of the samples. The SiC polytypes which form are not believed to be important from a ballistic perspective.[1,4]

The grain size varied by a factor of five, with AlN slowing down grain growth and Al and B₄C promoting grain growth. The aspect ratio shown in Table II is not the average of all grains, but rather the average of the most acicular grains in each sample. All materials were primarily equiaxed and there were not significant differences in their aspect ratios.

Young's modulus was highest for SiC-N and lowest for the SiC densified with boron and carbon. High Young's modulus was obtained by having high density and low amounts of secondary phases. Transmission electron microscopy showed that only the SiC densified with boron and carbon was devoid of triple points containing Al and O.[15] SiC-N had lower amounts of this secondary phase based on TEM observations of triple points and grain boundaries. The lower amount of O and N, as well as the inability to plasma etch the SiC-N microstructure support this same conclusion. The high Young's modulus of SiC-N is a distinguishing feature of the material. Poisson's ratio ranged from 0.183 (B-C) to 0.199 (1950°C Al).

The amount of retained O and N is dependent on processing conditions since O can be evolved from SiC in the form of SiO, CO, and Al₂O much easier when it has open porosity than when the porosity is closed. It is likely that the grain boundary chemistry in combination with the grain size influences the fracture mode. Figure 2 provides a view of the fracture mode of the six materials. The cracking during indenting, as shown in Figure 3, also gives insight into the fracture mode. The most detailed information was obtained by etching a polished, precracked SEPB bar and quantifying the fracture mode as shown in Table III. The SiC densified with B₄C and C has the lowest amount of intergranular fracture since grain boundaries are clean and there is no material at triple points to create residual stresses upon cooling. SiC-N and the silicon carbide densified with AlN had the most intergranular fracture. Small differences in processing conditions, however, can lead to large differences in the tendency to fracture transgranularly.[3] In fact, the SiC-2.5 wt. % AlN-0.5 wt. % B₄C was chosen for this study since it was expected to fracture primarily transgranularly[3], contrary to the results observed. The SiC-5 wt. % AlN, which previously fractured intergranularly[3] appears to fracture primarily transgranular fracture in Figure 2 but characterization of a precracked sample after etching showed that it was mostly intergranular (see Table III). This resulted in one material with primarily transgranular fracture, three with mixed-mode fracture, and two with mainly intergranular fracture for the ballistic evaluation.

**Table III**
**Mechanical Characterization**

| Designation | Strength (MPa) | | | % Intergranular | Toughness | Hardness (GPa) | |
|---|---|---|---|---|---|---|---|
| | Mean | Char.[a] | m[b] | Fracture | (MPa-m$^{1/2}$) | HK1 | HV1 |
| SiC-N | 563±73 | 597 | 8.5 | 77 | 4.5±0.1 | 20.0±0.2 | 21.9±0.4 |
| B-C | 343±60 | 367 | 6.7 | 6 | 2.5±0.1 | 20.8±0.2 | 23.7±0.5 |
| Al (1950°C) | 477±22 | 487 | 26.6 | 49 | 3.9±0.1 | 20.4±0.2 | 22.6±0.1 |
| Al (2200°C) | 529±67 | 563 | 7.9 | 59 | 4.9±0.2 | 19.0±0.3 | 21.0±0.7 |
| AlN | 617±143 | 673 | 4.9 | 75 | 3.1±0.1 | 20.9±0.2 | 25.5±0.7 |
| AlN-B₄C | 444±46 | 467 | 9.9 | 54 | 3.6±0.1 | 18.6±0.2 | 22.1±0.7 |

a. Characteristic strength (63.2 % probability of failure).
b. Weibull modulus.

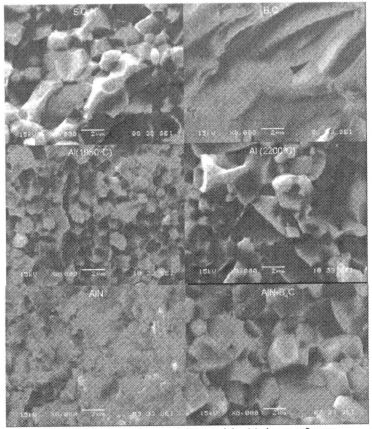

Figure 2. Fracture surfaces of six materials. Markers are 2 μm.

Table III also lists strength, fracture toughness, and hardness data. The fine grain size of the SiC-5 wt. % AlN gave it the highest Vickers hardness of any of the compositions. It is unclear why the SiC-2 wt. % C-0.5 wt. % B₄C had lower hardness, since this material tends to have a HV1 hardness in the range of 25-26 GPa.[1,6,16] The hardness-toughness tradeoff displayed by the SiC-1.65 wt. % Al samples is in accord with expectation.[3] The fracture toughness varied by a factor of two between materials (see Table III). While higher toughness is possible in SiC-based materials[1,3] the objective of this study was not to optimize toughness, but rather to fabricate materials which all had low porosity and high modulus, while still allowing a variation in toughness and hardness. The HV1 hardness-toughness range of the materials tested ballistically is shown in Figure 4. Knoop measurements compresses the hardness values, but a similar trend of decreasing hardness with increasing toughness is still observed.

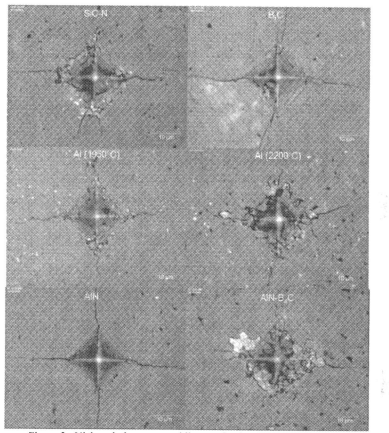

Figure 3. Vickers indents at one-kilogram loads. Markers are 10 μm.

Weibull plots for the bars tested in flexure are shown in Figure 5. The SiC-5 wt. % AlN had the highest strength, but also the lowest Weibull modulus. The SiC-1.65 wt. % Al material densified at 1950°C had moderate strength, but a very high Weibull modulus. Due to the large volume under stress in a tile subjected to a ballistic impact, one would expect that it would be very easy to find large flaws under stress to initiate failure compared to the flexure bars, which have a small effective volume.[17] Using this logic, the strength should be ranked according to a low probability of failure, which unfortunately requires an extrapolation of the data shown in Figure 5. The strength of the disks subjected to ballistic testing was highest for SiC-1.65 wt. % Al densified at 1950°C and likely lowest for SiC-2 wt. %C-0.5 wt. %B$_4$C since extrapolated Weibull plots tend to underestimate the strength at low probabilities of failure. The strength increases with increasing fracture toughness and decreasing grain size, as expected.[1]

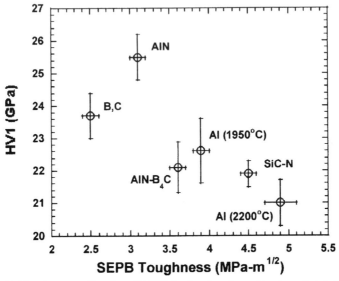

Figure 4. HV1 hardness as a function of SEPB fracture toughness for six materials tested.

Ballistic Testing

    The ballistic $V_{50}$ testing ranked the materials in the order shown in Table IV, with the first two materials higher than the middle two, which were significantly higher than the two materials listed last. The relative ballistic ranking was not predictable based solely on hardness, toughness, strength, grain size, elastic modulus, or fracture mode for this armor package at this threat level. Materials with higher Weibull values generally performed better than ceramics with lower Weibull moduli. The consistency of the microstructure or flaw population dictates

**Table IV**
**Ballistic Test Results Compared to Room Temperature Properties**

| $V_{50}$ Rank (high to low) | HV1 (GPa) | HK1 (GPa) | $K_{Ic}$ (MPa-m$^{1/2}$) | E (GPa) | Density (g/cc) | Strength (MPa) | m | % Intergranular Fracture |
|---|---|---|---|---|---|---|---|---|
| Al (1950°C) | 22.6 | 20.4 | 3.9 | 431 | 3.21 | 477 | 26.6 | 49 |
| AlN-B₄C | 22.1 | 18.6 | 3.6 | 423 | 3.21 | 444 | 9.9 | 54 |
| SiC-N | 21.9 | 20.0 | 4.5 | 442 | 3.22 | 563 | 8.5 | 77 |
| B₄C | 23.7 | 20.8 | 2.5 | 421 | 3.19 | 367 | 6.7 | 6 |
| AlN | 25.5 | 20.9 | 3.1 | 429 | 3.22 | 617 | 4.9 | 75 |
| Al (2200°C) | 21.0 | 19.0 | 4.9 | 428 | 3.21 | 529 | 7.9 | 59 |

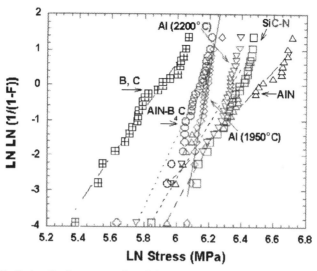

Figure 5. Weibull plots for flexure samples of six materials tested at room temperature.

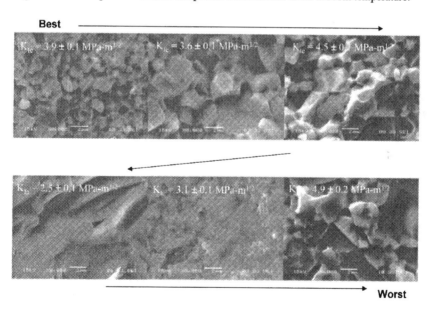

Figure 6. Comparison of microstructures with ballistic $V_{50}$ performance showing no link between grain size or fracture mode and ballistic ranking.

the Weibull modulus of a material. Reymann[18] concluded that materials for high-speed impact protection should have very small grain size and be as homogeneous as possible. While it is perfectly clear that fine grain size is not a predictor of $V_{50}$ performance at this threat level, a homogenous microstructure may be advantageous.[17] The role of crack initiation in flexure tests is very different from confined failure due to plastic yielding in ballistic events. It is generally expected that microstructure and fracture mode would be more important than strength. Lankford indicated that interlocked comminution could be dependent on grain size and fracture mode.[19] Figure 6, however, suggests that neither grain size nor fracture mode influenced performance. The tested debris was analyzed and fracture modes were similar to that seen in quasi-static fracture.

It is therefore interesting to calculate the flexural strength of the six materials by extrapolating their Weibull distributions to lower probabilities of failure as shown in Table V. As mentioned earlier, it is dangerous to assume that the distributions are valid far away from the data gathered. In order to get the ranking on the far right hand column of Table V, it would be necessary to test 10 million bend bars. While the $V_{50}$ data fit the strength data better than any other single parameter when extrapolated to low probabilities of failure, it is clear that the correlation is still poor. It is not surprising that no single parameter is able to predict ballistic performance.

It is very apparent that a number of materials perform well and many of the theories of what makes good ceramic armor do not hold up when compared to the data. For example, some have hypothesized that SiC-N is a good armor material since nitrogen diffuses into the SiC grains introducing localized strains which promote intergranular fracture. It is apparent that SiC can be made to have similar performance without introducing nitrogen at this threat level. Ironically, the best performing composition is very similar to the early work reported by Alliegro, et al. nearly 50 years ago.[20] The relatively good performance of the SiC-2 wt. % C-0.5 wt. % B$_4$C validates the claim of Lillo, et al. that a well-densified solid-state pressureless sintered SiC has a ballistic performance only slightly worse than SiC-N at moderate threat levels.[21] A TEM comparison of the materials tested showed that the SiC-2 wt. % C-0.5 wt. % B$_4$C composition was the only material with clean grain boundaries and triple junctions, while all of the other five compositions had Al and O at triple junctions. The triple point chemistry was surprisingly similar for all five materials despite that some had N and B in addition to Al and O. The SiC-N material had less of this secondary phase with many triple junctions devoid of Al and O. It is interesting to note that SiC-5 % AlN, one of the worst performing materials, had the highest amount of secondary phase present. It is unclear, however, what role the amount of

## Table V
### Flexural Strength at Different Failure Probabilities

| $V_{50}$ Rank (high to low) | F=0.5 | F=$10^{-1}$ | F=$10^{-2}$ | F=$10^{-3}$ | F=$10^{-4}$ | F=$10^{-5}$ | F=$10^{-6}$ | F=$10^{-7}$ |
|---|---|---|---|---|---|---|---|---|
| Al (1950°C) | 480 | 447 | 409 | 376 | 344 | 316 | 290 | 266 |
| AlN-B$_4$C | 451 | 373 | 294 | 233 | 185 | 147 | 116 | 92 |
| SiC-N | 571 | 458 | 348 | 265 | 202 | 154 | 118 | 90 |
| B$_4$C | 348 | 263 | 185 | 131 | 93 | 66 | 47 | 33 |
| AlN | 625 | 426 | 265 | 166 | 104 | 65 | 41 | 26 |
| Al (2200°C) | 537 | 424 | 315 | 235 | 176 | 131 | 98 | 73 |

secondary phase plays in such a situation since the two SiC-1.65 wt. % Al materials behaved differently. The highest partial for the material densified at 1950°C was 50 m/s higher than the highest partial for the same composition densified at 2200°C.

This testing suggests that a variety of materials can perform acceptably at this threat level and that SiC with small additions of Al (1.65 wt. %) can be densified at relatively low temperatures (1950°C) for short times (1 hour) at reasonable pressures (21 MPa) is one of these materials. Optimization of such a material would be worthwhile if it can reduce the cost of hot pressed armor. A greater need, however, is to make less expensive armor by pressureless sintering, since large tanks take large amounts of material. This is a greater challenge since the absence of applied pressure makes the attainment of high density and microstructural engineering more difficult.

## CONCLUSIONS

A SiC-1.65 wt. % Al hot pressed at 1950°C and a SiC-2.5 wt. % AlN-0.5 wt. % $B_4C$ had comparable or slightly better $V_{50}$ performance as SiC-N. SiC-5 wt. % AlN and SiC-1.65 wt. % Al hot pressed at 2200°C had markedly worse $V_{50}$ performance under the same test conditions. A SiC-2 wt. % C-0.5 wt. % $B_4C$ material was intermediate in performance. The $V_{50}$ ranking was independent of the hardness, fracture toughness, mean strength fracture mode, density, Young's modulus, grain size, retained oxygen and nitrogen, and polytypes present. The materials which performed best had higher Weibull moduli than the other three samples. When the strength data were fitted with a two-parameter Weibull distribution and extrapolated to low probabilities of failure, as would be expected for ballistic impact, the strength was in fair agreement with the $V_{50}$ data. It is not expected, however, that this strength is a predictive parameter since the correlation with $V_{50}$ data was still poor.

This testing, however, showed that differences in ballistic performance do exist between materials with similar density, modulus, and composition. While microstructural differences do not appear to affect the ballistic performance in a simple manner, it is worthwhile to continue to make comparative samples in an effort to understand what controls the ballistic events. In one sense, one result of this work is evidence that some of the hypotheses about what makes good armor were proved false.

The need exists to lower the cost of ceramic armor in order to allow mass production to occur. While the hot pressed SiC-Al material which performed best in this study is easy to manufacture, it is still more expensive than a pressureless sintered product. Efforts should be directed towards understanding differences in pressureless sintered SiC armor with densities greater than 99 % of theoretical.

## ACKNOWLEDGEMENT

This SBIR work was performed for the U.S. Army under contract DAAD17-02-C-0052. Appreciation is expressed to Lyle Miller of Ceramatec for help with x-ray diffraction work and Sarbjit Kaur for microstructural characterization. Technical discussions with Dr.William Rafaniello and Dr. Svante Prochazka are gratefully acknowledged.

## REFERENCES
[1] D. Ray, R. M. Flinders, A. Anderson, R. A. Cutler and W. Rafaniello, "Effect of Room-Temperature Hardness and Toughness on the Ballistic Performance of SiC-Based Ceramics," *Ceram. Eng. Sci. Proc.*, **26**[7], (2005).

[2]A. Krell and E. Strassburger, "High Purity Submicron α-Al₂O₃ Armor Ceramics: Design, Manufacture, and Ballistic Performance," *Ceram. Trans.*, **134**, 463-71 (2002).

[3] R. Marc Flinders, D. Ray, A. Anderson and R. A. Cutler, "Microstructural Engineering of the Si-C-Al-O-N System," *Ceram. Trans.* **178**, 63-78 (2005).

[4]R. Marc Flinders, D. Ray, A. Anderson and R. A. Cutler, "High-Toughness Silicon Carbide as Armor," *J. Am. Ceram. Soc.* **88**[8], 2217-26 (2005).

[5]ASTM C 1421-99, Standard Test Methods for Determination of Fracture Toughness of Advanced Ceramics at Ambient Temperature, pp. 641-672 in *1999 Annual Book of Standards* (ASTM, Philadelphia, PA 1999).

[6]D. Ray, M. Flinders, A. Anderson, and R. A. Cutler, "Hardness/Toughness Relationship for SiC Armor," *Ceram. Sci. and Eng. Proc.*, **24**, 401-10 (2003).

[7]H. M. Rietveld, "A Profile Refinement Method in Neutron and Magnetic Structures," *J. Appl. Crystallogr.*, **2**, 65-71 (1969).

[8]D. L. Bish and S. A. Howard, "Quantitative Phase Analysis Using the Rietveld Method," *J. Appl. Crystallogr.*, **21**, 86-91 (1988).

[9]D. H. Stutz, S. Prochazka and J. Lorenz, "Sintering and Microstructure Formation of b-Silicon Carbide," *J. Am. Ceram. Soc.*, **68**[9], 479-82 (1985).

[10]E. E. Underwood, *Quantitative Stereology*, (Addison-Wesley, Reading, MA. 1970).

[11]R. A. Cutler and T. B. Jackson, 'Liquid Phase Sintered Silicon Carbide," pp. 309-318A in *Third International Symposium on Ceramic Materials and Components for Engines,*" ed. by V. J. Tennery (Am. Ceram. Soc., Westerville, OH. 1989).

[12]I. B. Cutler, P. D. Miller, W. Rafaniello, H. K. Park, D. P. Thompson and K. H. Jack, "New Materials in the Si-C-Al-O-N and Related Systems," *Nature*, **275**, 434-35 (1978).

[13]A. Zangvil and R. Ruh, "Phase Relationships in the Silicon Carbide-Aluminum Nitride System," *J. Am. Ceram. Soc.*, **71**[10], 884-90 (1988).

[14]R. M. Williams, B. N. Juterbock, S. S. Shinozaki, C. R. Peters, and T. J. Whalen, "Effects of Sintering Temperatures on the Physical and Crystallographic Properties of β-Silicon Carbide," *Am. Ceram. Soc. Bull.*, **64**[10], 1385-9 (1985).

[15]E. Pabit, S. Crane, K. Siebein, D. P. Butt, D. Ray, M. Flinders, and R. A. Cutler, "Grain Boundary and Triple Junction Chemistry of Silicon Carbide with Aluminum or Aluminum Nitride Additive," *Ceram. Trans.* **178**, 91-102 (2005).

[16]M. Flinders, D. Ray and R. A. Cutler, "Toughness-Hardness Trade-Off in Advanced SiC Armor," *Ceram. Trans.*, **151**, 37-48 (2003).

[17]M. P. Bakas, V. A. Greenhut, D. E. Niesz, G. D. Quinn, J. W. McCauley, A. A. Wereszczak, and J. J. Swab, "Anomolous Defects and Dynamic Failure of Armor Ceramics, *Int. J. Appl. Ceram. Technol.*, **1**[3], 211-18 (2004).

[18]J. J. Reymann, "Characterization of Ceramics Subjected to Dynamic Stresses," *Materiaux Techniques*, **73**[4-5], 133-40 (1985).

[19]J. Lankford, Jr., "The Role of Dynamic Material Properties in the Performance of Ceramic Armor," *Int. J. Appl. Ceram. Technol.*, **1**[3], 205-10 (2004).

[20]R. A. Alliegro, L. B. Coffin, and J. R. Tinklepaugh, "Pressure-Sintered Silicon Carbide," *J. Am. Ceram. Soc.*, **39**[11], 386-89 (1956).

[21]T. M. Lillo, H. S. Chu, D. W. Bailey, W. M. Harrison, and D. A. Laughton, "Development of a Pressureless Sintered Silicon Carbide Monolith and Special-Shaped Silicon Carbide Whisker-Reinforced Silicon Carbide Matrix Composite for Lightweight Armor Applications," *Ceram. Eng. Sci. Proc.*, **24**[3], 359-64 (2003).

# ADDITION OF EXCESS CARBON TO SIC TO STUDY ITS EFFECT ON SILICON CARBIDE (SIC) ARMOR

Chris Ziccardi
Rutgers University
607 Taylor Road
Piscataway, NJ, 08854-8065

Richard Haber
Rutgers University
607 Taylor Road
Piscataway, NJ, 08854-8065

ABSTRACT:

Commercial silicon carbide (SiC) powders are composed of a wide range of particle sizes, typically 0.5 to 15um depending on the powder. It was shown that by eliminating large particles and hard agglomerates greater than 2um improved microstructural uniformity in hot pressed parts could be achieved. The effect of carbon over-additions ranging from 0.25 to 5 % on microstructural uniformity was also examined. It was found that for over-additions of carbon exceed 1% the Knoop hardness was shown to decrease by as much as 20% compared with conventionally prepared, dense SiC. The effect of increased carbon percentages on microstructure and hardness was shown.

INTRODUCTION:

Current SiC armor materials exhibit great variability that is believed to be due to variations in microstructure. In a study by Bakas et.al. on rubble from ballistically impacted hot pressed SiC, it was found that a variety of defects are commonly found on the fracture surfaces. A common defect polytype found included carbonaceous particulates. The origin of these defects was not discussed, however it was speculated that these inclusions were related to the carbon batch additions. They were able to show that the percentage and size of these defects had an effect on ballistic performance.[1]

When hot pressing SiC, two types of additives are commonly used. The first is carbon and the second include a variety of metallic oxide, carbides or nitrides. In the case of the later, boron, aluminum and yttrium are most often used.[2,3] The exact carbon and cation source can vary among manufacturers. The carbon is added to remove the silica passivation layer present on the SiC powder particles by thermodynamic reaction of C + $SiO_2$ by vapor transport of $SiO(g)$ and $CO(g)$ within the pores of the material prior to densification. Without the removal of this silica layer, the resultant microstructure would undergo excessive coarsening, resulting in incomplete densification. Simulation studies by Kaza et. al. have shown that for a 2um SiC powder which used in this study, 0.8% total carbon is required to eliminate the silica layer. The HX4 additive provided by Huntsman provides the 0.8% carbon necessary to remove the oxide layer in this specific powder. In industry however it is common to see additions of carbon sources, ie. phenolic resin or graphite additions in excess of 3%.[4,5]

The presence of the cationic additive affects the sintering behavior of the SiC by either creating a liquid which aids in densification or by inhibiting surface diffusion and thereby allowing bulk diffusion to dominate sintering. In the case of solid state additives such as boron, Rafaniello and Ness showed that the boron provides a reduction in surface diffusion resulting in a finer particle size, dense microstructure. The cationic additives are commonly made at 0.7-1.5% levels.

Ziccardi and Haber[6] showed that an aqueous based surfactant could be used as the major source of both boron and carbon. The SiC microstructures were dense and fine grain size using this approach.

The effect of over additions of additives has not been discussed in the literature; however, it is reasonable to believe that higher percentages of either carbonaceous or cationic additives will lead to the presence of additional phases within the microstructures. The effect of these secondary phases on ballistic performance is most likely detrimental. This study will examine the effect of small excess additions of carbon on microstructure and physical properties.

EXPERIMENTAL AND DISCUSSION:

The composition used as the basis for this study was similar to what was reported by Ziccardi and Haber.[6] This base formulation used both carbon and boron based additives. The starting SiC powder was a 2-4um, alpha SiC from UK Abrasives (Northbrook, IL). A proprietary, aqueous based surfactant HX4 (Huntsman HX4 distributed by H.C. Spinks Co, Paris, TN) was used as the primary boron source. A UK Abrasive's 2 um, boron carbide powder was used as a secondary boron source. The HX4 additive provided a sufficient amount of carbon to eliminate the silica layer on the starting powders. Excess additions of carbon were made by incorporating from 0% to 5% into the batch,.

The SiC powder along with 3.5% HX4 and 0.5% of boron carbide were dry mixed in a ball mill for 2 hours. Future studies samples will be wet mixed to determine the effect mixing has on both the size and presence of these carbonaceous defects. For the samples with higher percentages of carbon, a 3-5um size graphitic powder from Industrial Graphite Sales LLC (Elkgrove, IL) was mixed to the base batch and similarly mixed. 0.25 to 5% over additions of carbon were examined. Mixed batch was then dry pressed to 250 psi (1.7 Mpa) in a uniaxial die 2.25 inches in diameter. The formed green disks were densified by hot pressing. Dried disks were loaded in a graphite die. Using a Vacuum Industries Model 4-2078 hot press, samples were heated at a rate of 25° C per hour to 1500°C under vacuum. The sample was held for one hour at 1500°C. This dwell was added to allow the carbon present from the surfactant to reduce the native $SiO_2$ layer present on the surface of the SiC powder. After one hour, the sample was heated to 1800°C where the hot press was then backfilled with argon and the sample pressed to 5.6 Ksi (38 MPa). The sample was then heated to 2300°C for 15 minutes. The power was immediately turned off and the sample allowed to cool.[6]

RESULTS:

Microstructures were evaluated using a field emission scanning electron microscope. Specimens were obtained from static flexure test bars. Figure 1 the microstructure of the base SiC formulation. No additions of graphite were made.

Figure 1.  Microstructure of the base SiC formulation in this study.

Figures 2-6 show the fracture surfaces for over-additions of carbon made to the base formulations.  With increasing carbon additions, carbon particulates become increasingly evident on the surfaces. For over-additions greater than 2%, the particulate inclusions are tend to be larger, most likely due to carbon agglomerates. These inclusions appear to be somewhat porous. The morphology of these carbonaceous inclusions closely resembles those observed by Bakas et.al in his studies.

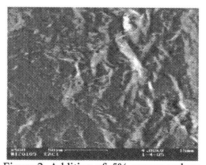

Figure 2. Addition of .5% excess carbon.

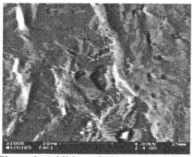

Figure 3. Addition of 1% excess carbon.

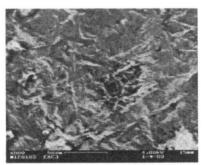
Figure 4. Addition of 2% excess carbon.

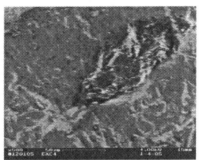
Figure 5. Addition of 4% excess carbon.

Figure 6. Addition of 5% excess carbon.

Knoop hardness data was collected for each sample. Figure 7 compares the hardness versus load profile for the base formulation through 5% carbon over-additions. In each case, hardness measurements were made for a range of loads, 250g to 2 Kg. A SRM2830 NIST standard reference block was used to ensure the accuracy of the measurements. Hardness measurements were made on flat surfaces, polished to 1 um. When making the indents, the regions with carbon inclusions were avoided. This could conclude that only the hardness to density drop off ratio is being evaluated.

Density started from 3.206 at 0% carbon and moved down to 3.193, 3.187, 3.174, 3.152, and 3.143 for .5% carbon, 1% carbon, 2% carbon, 4% carbon, and 5% carbon respectively. However, scattering in the Weibull data made it seem as if there was more going on than just a density change. Each point represented the average of ten indents. The results show a consistent trend in reduced hardness with increasing carbon content. The widest spread in hardness values was shown at 1 Kg. This spread decreased with increasing load. The 1 Kg load indents provided a sufficiently large sampling area to differentiate between samples. The higher load exhibited cracking and was more representative of the bulk SiC. Figure 8 showed the statistical analysis of this data. Using a Weibull analysis, it is evident that with increasing carbon over-additions, there is an increased variance in the hardness data. The microstructural uniformity of the SiC decreases with increased carbon content. This data is significant in that it may provide insight as to why some commercial SiC materials have more variable physical and ballistic properties. Over-additions or poor mixedness of additives will result in less uniform microstructures and more variable properties.

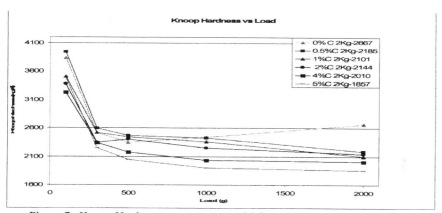

Figure 7. Knoop Hardness measurements with increasing percentage of carbon

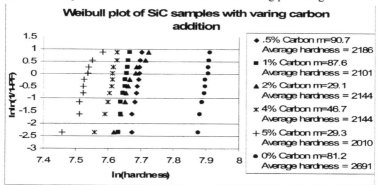

Figure 8. Weibull plot with increasing percent of carbon the data was calculated a 2Kg load.

Figure 9 compares a single carbonaceous inclusion found in this study with an inclusion reported by Bakas et.al. The similarity in morphology is evident. This type of layered inclusion is most likely the result of a poorly mixed carbon additives compacted and flattened during hot pressing. However, those observed by Bakas et.al. were in significantly lesser volumes than observed in this study.

Figure 9. Potato chip defects, on the left, found in this study, and on the right was found in the Bakas and Greenhut work. [1]

REFERANCES:
(1) M.P Bakas, V.A. Greenhut, D.E Niesz, G.D. Quinn, J.W. McCauley, A.A Wereszczak, and J.J. Swab, "Anomalous Defects and Dynamic Material Properties in the Performance," International Journal of Applied Ceramic Technology, Vol.1[3] 211(2004).

(2) E. Ness and W. Rafaniello, "Origin of Density Gradients in Sintered β–Silicon Carbide Parts," J. Am. Ceram. Soc., 77 [11] 2879-84(1994).

(3) J. Ihle, M. Herrmann, J. Adler, "Phase formation in porous liquid phase sintered silicon carbide: Part I: Interaction between Al2O3 and SiC," Journal of the European Ceramic Society 25 (2005) 1005–1013.

(4) W. Van Rijswiik and D. J. Shanefield, "Effects of Carbon as a Sintering Aid in Silicon Carbide," J. Am. Ceram. Soc., 73 [1] 148-49(1990).

(5) A. Kaza, M. J. Matthewson and D. E. Niesz, "Removal of SiO2 from green SiC compacts", 106th Annual Meeting of the American Ceramic Society, Indianapolis, April 2004

(6) C. Ziccardi, V. Demirbas, R. Haber, D. Niesz, and J. McCauley, "Means of Using Advance Processing to Eliminate Anomalous Defects on SiC Armor," Advances in Ceramic Armor (Proceedings of the 29th International Conference on Advanced

Cermanic and Compostitic Cocoa Beach Florida USA 2005), Novel Material Concepts 271-277.

(7) De Angelis, C. Rizzo, E. Ferretti, and S.P. Howlett, "Sintering of –SiC: Dispersion of the Carbon Sintering Aid," Ceramics Today – Tomorrow's Ceramics, 1415-1423(1991).

# Glass and
# Transparent Ceramics

# ANALYSIS OF TIME-RESOLVED PENETRATION OF LONG RODS INTO GLASS TARGETS—II

Charles E. Anderson, Jr., I. Sidney Chocron, and Carl E. Weiss
Southwest Research Institute
P.O. Drawer 28510
San Antonio, TX 78228-0510

## ABSTRACT

The penetration response of tungsten-alloy long-rod penetrators into soda-lime glass targets was reported at the last Cocoa Beach meeting [1]. Numerical simulations using a Drucker-Prager constitutive model were compared to experimental data at two impact velocities: 1.25 km/s and 1.70 km/s. Preliminary values for the slope and cap of the Drucker-Prager model were selected. In the current study, a modified Drucker-Prager model is examined. Results of numerical parametric studies are compared to the experimental data to determine optimal values for the constitutive parameters. At the end, it is concluded that numerical issues may be obfuscating the mechanics.

## INTRODUCTION

Anderson, *et al.*, revisited experiments of the impact of tungsten-alloy, long-rod projectiles into glass targets in Ref. [1]. Interest in glass has increased because of the observation of the phenomenon referred to as failure waves, and because of the importance of transparent "armor" for vehicles to provide adequate ballistic protection against a variety of threats. In addition, computational power has increased tremendously over the last 12 or so years. Therefore, it was decided to re-examine the work conducted earlier [2-4]. A brief summary of the findings of Ref. [1] will be provided after a description of the experimental results is given; and then, the results of parametric studies conducted to assist in determination of the parameters for a modified Drucker-Prager constitutive model are summarized.

## THE EXPERIMENTS

For completeness, the details of the experiment are repeated here in this paper. The experiments were conducted by Hohler and Stilp at the Fraunhofer Institut Kurzzeitdynamik (Ernst-Mach-Institut) [2-3]. The blunt-nose, tungsten-alloy (density 17.6 g/cm$^3$) projectiles had a length-to-diameter ratio (*L/D*) of 12.5, and a diameter of 5.8 mm. The targets were fabricated from sheets of soda-lime (float) glass, density 2.5 g/cm$^3$, with a compressive strength of 0.90 GPa. Ten sheets of 1.9-cm-thick glass and one sheet of 1.0-cm-thick glass were bonded together using double-sided adhesive (0.014-cm thick), for a total thickness of 20.0 cm. The glass was backed by a block of mild steel, as shown in Fig. 1. The lateral dimensions for most of the glass targets were 15 cm by 15 cm. Two of the experiments were performed with 30-cm x 30-cm wide glass plates to investigate if there was any influence of the lateral dimensions. The experimental results were independent of this doubling of the lateral dimensions.

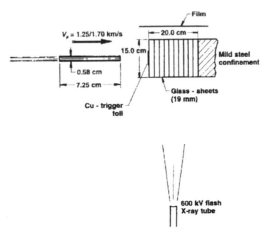

Fig. 1. Schematic of experimental arrangement.

Flash radiography was used to record the positions of the front and tail of the projectile, as well as the length of the projectile, as a function of time after impact. The nominal impact velocities for two sets of experiments were 1.25 km/s and 1.70 km/s.

The position-time of the nose and tail of the rods are shown in Fig. 2 for the two sets of experiments. Each pair of points (nose and tail positions) represents one experiment. The experimental spread in the impact velocities was ±20 m/s. The experimental data were adjusted by a simple proportionality to bring all data to a common impact velocity of 1.25 or 1.70 km/s. In none of the experiments did the projectile penetrate to the mild steel block.

A modified target was constructed for some of the tests to provide an estimate of the projectile velocity at different depths of penetration. This experimental configuration is shown in Fig. 3. The glass portion of the target had different thicknesses, and the projectile was allowed to exit the target through a 3.0-cm wide cavity drilled into the mild steel backup plate. A 4.5-cm wide, 1.5-cm thick, honeycomb sheet was placed at the rear of the glass plates to provide confinement but minimize penetration/perforation resistance. Flash radiography was then used to obtain the residual velocity and projectile length behind the target.

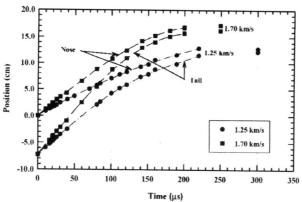

Fig. 2. Position-time data for experiments into glass. The dashed lines represent regression fits through the data.

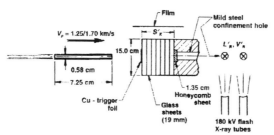

Fig. 3. Schematic of modified target arrangement (600 kV X-ray—not shown in this figure—is the same as Fig. 1).

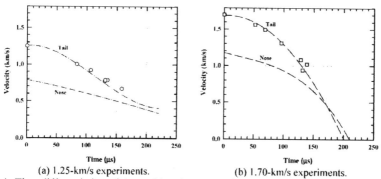

(a) 1.25-km/s experiments.  (b) 1.70-km/s experiments.

Fig. 4. Time differentiation of the position-time regression fits in Fig. 2 give the tail and nose (penetration) velocities vs. time.

Least-squares polynomial regression fits to the position-time data are shown as the dashed lines in Fig. 2. Differentiation of these regression fits with respect to time gives the tail and nose (penetration) velocities vs. time for the two sets of data. These are plotted in Fig. 4. Also plotted in the figures are the measured rod velocities using the experimental arrangement of Fig. 3. The residual rods nominally have the same velocity as the tail since most of the projectile is moving with the tail velocity [5].

The normalized lengths of the rod vs. time are shown in Fig. 5. The closed symbols denote data using the experimental arrangement shown in Fig. 1, while the open symbols denote data using the arrangement shown in Fig. 3 (after the rods have left the target). Additionally, some of the residual rods were recovered in the impact tank; these are plotted as open symbols with a center dot beyond 300 μs. The transition from eroding to rigid-body penetration occurs around 130 μs - 140 μs for both impact velocities. Thus, penetration into glass can be divided into two distinct phases: eroding penetration and rigid-body penetration.

NUMERICAL SIMULATIONS

CTH [6] was used for the numerical simulations. Twenty-five zones were used to resolve the radius of the projectile. With this resolution, the treatment of mixed cells should have less than a 4% effect on the penetration results; thus, avoiding the problem observed in Ref. [2]. It was assumed that the projectile penetrated failed glass, which could be described by a Drucker-Prager constitutive model:

$$Y = \min(\beta P, \overline{Y})$$

(1)

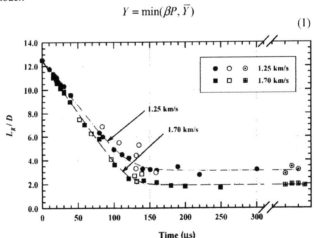

Fig. 5. Normalized residual rod length vs. time.

where $Y$ is the flow stress and $P$ is the hydrostatic pressure. The pressure-dependent region corresponds to comminuted pieces sliding over each other; and the cutoff, or cap, corresponds to material-deforming flow. The model has two adjustable constants, $\beta$ and $\overline{Y}$: the slope of a pressure-dependent region, and a cap that limits the flow stress. It was determined that the response at 1.70 km/s is controlled largely by the cap, and the response at 1.25 km/s is controlled by the slope. One set of constants for both impact velocities ($\beta = 2.0$ and $\overline{Y} = 1.5$ GPa) provided approximate agreement between the simulations and experimental data.

The objective of the current work was to conduct parametric simulations to optimize the values of $\beta$ and $\overline{Y}$. However, a convergence study was first conducted, and it was determined that 7.5 zones across the radius gave the same results as 25 zones, with a substantial speed up in execution time. Thus, all subsequent simulations were conducted using 7.5 zones to resolve the radius of the projectile.

A large number of simulations were conducted in an attempt to optimize $\beta$ and $\overline{Y}$, comparing the results of nose and tail positions to the experimental data at the two impact velocities. Concurrently, experiments were being conducted on failed glass [7-8], and these provided insights to possible values for Eqn (1). It was found that the fractured glass has some intrinsic strength at zero axial load [8]. This implies that failed glass has tensile carrying capabilities under confinement. Thus, the failed glass can develop tensile hoop stresses in the divergent stress field in front of the projectile during penetration. The ability of comminuted material to support tensile stresses, at first, seems counterintuitive; but, the comminuted glass particles are interlocked, and cannot necessarily move (apart) with an applied tensile stress. Additionally, as the pressure decreases, so does $Y$ in Eqn. (1); and thus, the resistance to penetration also decreases. Without a finite, non-zero value for $Y_0$, penetration will continue forever, albeit at lower and lower penetration velocities. In the experiments, the remnant projectile never penetrated to the steel backup plate. Thus, an alternative form of the Drucker-Prager model was adopted:

$$Y = \min(\beta P, \overline{Y}) + Y_o \qquad (2)$$

where $Y_0$ provides a minimum resistance to penetration.

Preliminary analysis of data [8] suggests that $\beta$ should be approximately 1.6. Therefore, for the initial computations, $\beta$ was set to 1.6, and $Y_0$ was set to zero. Since we had determined previously that the cap played a very minor role at 1.25 km/s, we focused on the data at the lower impact velocities (and set the cap effectively to infinity). The numerical simulation was in reasonable agreement for the penetration-time response, but greatly underestimated the erosion of the projectile. The erosion rate strongly depends upon the strength of the projectile [5]. The constitutive form used for the tungsten alloy has the form [5]:

$$\sigma_W = Y_W[1 + 0.06\ln\dot\varepsilon] \qquad (3)$$

where $\dot\varepsilon$ is the strain rate. A parametric study was conducted where the initial flow stress, $Y_W$, of the tungsten-alloy rod was varied from 1.2 GPa to 2.0 GPa. The results are plotted in Figs. 6 and 7.

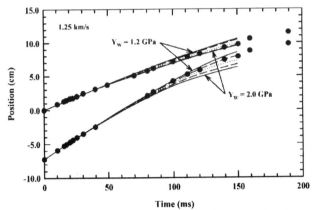

Fig. 6. Parametric study of projectile strength, where the flow stress $Y_W$ of the rod was varied from 1.2 GPa to 2.0 GPa in increments of 0.2 GPa ($Y_o = 0$).

With the results of Figs. 6 and 7, it was concluded that a $Y_W$ of 1.5 GPa appeared to be a reasonable value as a compromise of giving the correct penetration and rod length vs. time. A value of 1.5 GPa is low for current tungsten alloy materials, where a value of 2.0 GPa is generally more appropriate. However, the experiments were done circa the early 1980's, just about the time there was a significant effort to improve the strength and toughness of tungsten alloys; thus, a value of 1.5 GPa is probably okay.

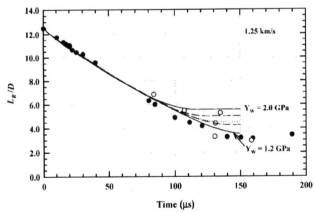

Fig. 7. Parametric study of projectile strength, where the flow stress $Y_W$ of the rod was varied from 1.2 GPa to 2.0 GPa in increments of 0.2 GPa ($Y_o = 0$).

The next parametric study examined the influence of $Y_o$. As can be seen in Fig. 8, $Y_o$ has a rather profound influence on the length of the rod. Although not shown, the penetration-time response is relatively unaffected for relatively small values of $Y_0$ until the later times, where "small" means 0.01 – 0.02 GPa. This is because the penetration pressures are considerably larger, compared to $Y_0$, at the earlier times.

It had now been determined that, for the conditions of no cap, that is, $\overline{Y} = 0$. and $Y_0$ approximately equal to 0.01-0.02 GPa, that the numerical simulations reasonably reproduce the experimental penetration-time results, including the rod length vs. time at 1.25 km/s. It is noted, for all cases, that the simulations underpredict the rod length for approximately the first 30-40 μs. We will return to this observation later.

We now conducted simulations of the 1.70-km/s experiments. The results for the rod length vs. time are shown in Fig. 9. Clearly, with no cap ($\overline{Y}$ = inf ), the erosion rate is much too high. A parametric study was done varying $\overline{Y}$, with values of 3.0, 2.0, and 1.0 GPa. After examining the results, it was concluded that a value of 2.0 GPa was too high, and a value of 1.0 GPa was too low, so one additional simulation was conducted with $\overline{Y}$ =1.5 GPa. This result is also shown in Fig. 9; the comparison is quite good.

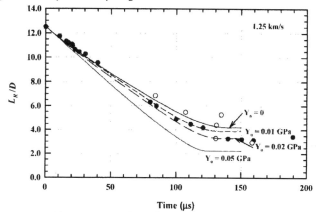

Fig. 8. Parametric study of $Y_0$ for $Y_W$ = 1.5 GPa.

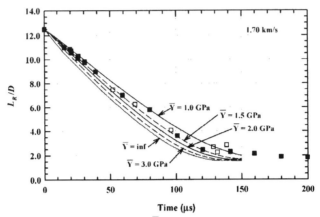

Fig. 9. Parametric study of $\overline{Y}$ for the 1.70-km/s experiments, with $\beta = 1.6$, $Y_0 = 0.01$ GPa.

The next parametric study examined the effect of $\beta$ at 1.70 km/s. The results are shown in Fig. 10. As can be seen, within the range of $\beta$ investigated, there is little sensitivity to the value of $\beta$, particularly early in the penetration process where the penetration pressures are sufficiently high that comminuted glass is on the cap. As the projectile decelerates, the pressure drops, and the effect of $\beta$ is observed.

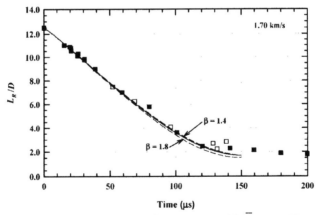

Fig. 10. Parametric study of $\beta$ at 1.70 km/s, with $\overline{Y} = 1.5$ GPa; $Y_0 = 0.015$ GPa ($\beta = 1.4$ & 1.8); $Y_0 = 0.010$ GPa ($\beta = 1.6$)

The effect of $\beta$ on simulated results at 1.25 km/s is shown in Fig. 11. The dotted line is for $\beta = 1.6$, as shown in Fig. 8 with $Y_0 = 0.01$ GPa and $\overline{Y} = \infty$. The other two simulations, for $\beta = 1.4$

and 1.8, have $\overline{Y} = 1.5$ GPa (and a $Y_0 = 0.015$ GPa). Note that the two simulations with the cap pass through the early-time data points. Although the cap is not particularly important for the 1.5-km/s-impact case, the cap does limit the stresses of the impact shock, just enough to alter the very early-time penetration and erosion rates.

Nominally optimized values have now been determined for the three parameters in the Drucker-Prager constitutive model of Eqn. (2): $\beta = 1.6$, $\overline{Y} = 1.5$ GPa, and $Y_0 = 0.015$ GPa. The results of simulations at the two impact velocities are compared to the experimental data in Fig. 12 (position vs. time) and Fig. 13 (normalized length vs. time). Overall, the comparison is quite good for the first ~130 μs, the time duration of eroding penetration. Rod length is slightly underpredicted at 1.25 km/s, and slightly overpredicted at 1.70 km/s. After 130-150 μs, the rod penetrates the failed glass in a rigid-body mode. It appears, from Fig. 12, that the penetration resistance for times greater than 130-150 μs is a little too large; thus, the strength of glass needs to be reduced somewhat.

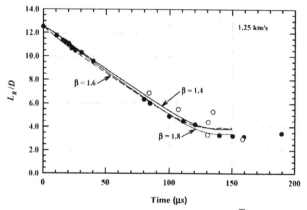

Fig. 11. Parametric study of $\beta$ at 1.25 km/s, with $\overline{Y} = 1.5$ GPa:
$Y_0 = 0.015$ GPa ($\beta = 1.4$ & 1.8); $Y_0 = 0.010$ GPa ($\beta = 1.6$)

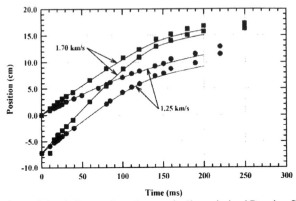

Fig. 12. Comparison of simulation results using nominally optimized Drucker-Prager parameters to experimental data for position vs. time.

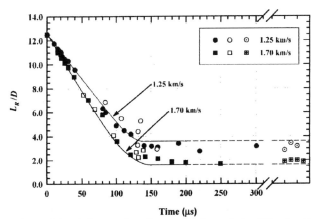

Fig. 13. Comparison of simulation results using nominally optimized Drucker-Prager parameters to experimental data for normalized length vs. time.

There is concern, however, with this computational study. Do the constitutive parameters determined by matching the experimental data truly describe the constitutive response of the material, or has this been an exercise in "numerology"? Of particular concern is the strong dependence of the simulated results on very small changes in $Y_o$. The value for $Y_o$ (~0.015 GPa, i.e., ~2 ksi) is quite small, and as such, should have little influence on penetration resistance. We examined pressure contours, and found that there are regions within the glass target where the pressure is negative (in tension) for the case when $Y_o$ is zero (which means the glass cannot support any shear); whereas, with a nonzero $Y_o$, the pressure contours are positive and considerably "smoother."

We had noticed previously [1] that the penetration velocities and pressures are very "noisy" for relatively large values of $\beta$, i.e., $\beta \geq 1.5$. Near the projectile-target interface, the penetration pressure is large, and because $\beta$ is large, there is a large equivalent stress (from the Drucker-Prager model), which implies a large penetration resistance. A large penetration resistance results in a decreased value of the penetration velocity. But when the penetration velocity decreases, there is an attendant drop in the pressure (since the pressure is proportional to the square of the penetration velocity). This drops the equivalent stress for the next computational cycle. Since the penetration resistance is now lower, the penetration velocity increases, thereby increasing the pressure. That is, the glass constitutive model is highly sensitive to small changes in pressure because a large $\beta$ acts as an "amplifier" of the pressure. (This is not such an issue for the 170-km/s-impact case since much of the target material is on the cap, which limits the flow stress, resulting in a relatively "smooth" penetration velocity.)

SUMMARY AND CONCLUSIONS

Constitutive parameters for soda-lime glass, assuming a modified Drucker-Prager model, were determined by conducting parametric studies on the para-meters and comparing the results to experimental data at two impact velocities. A set of constitutive parameters was found that do a fairly good job at reproducing the experimental data. Nevertheless, there is concern that numerical issues—in particular, large numerical oscillations—are "swamping" the mechanics. Also troublesome is that the current study determined a cap of 1.5 GPa, whereas on-going laboratory characterization of glass suggests that the cap should be higher [8]. Thus, the current work must be considered "work in progress" as we continue our investigations in characterization and numerical modeling.

REFERENCES

[1]C. E. Anderson, Jr., I. S. Chocron, and J.D. Walker, "Analysis of time-resolved penetration into glass targets," *Ceramic Engng. & Sci. Proc.*, **26**(7), *Proc. 29th Int. Conf. Advanced Ceramics & Composites*, 27-34, American Ceramics Society, 2005.

[2]C. E. Anderson, Jr., V. Hohler, J. D. Walker, and A. J. Stilp, "Penetration of long rods into steel and glass targets: experiments and computations," *Proc. 14th Int. Symp. on Ballistics*, Québec City, Canada, 24-29 September 1993.

[3]C. E. Anderson, Jr., V. Hohler, J. D. Walker, and A. J. Stilp, "Modeling long-rod penetration into glass targets," *14th U. S. Army Symp. on Solid Mech.*, (K. R Iyer and S-C Chou, Eds.), pp. 129-136, Battelle Press, Columbus, OH, 1996.

[4]J. D. Walker and C. E. Anderson, Jr., "An analytic penetration model for a Drucker-Prager yield surface with cutoff," *Shock Compression of Condensed Matter—1997*, S. C. Schmidt, Eds., 897-900, 1998.

[5]C. E. Anderson, Jr. and J. D. Walker, "An examination of long-rod penetration," *Int. J. Impact Engng.*, **11**(4), 481-501, 1991.

[6]J. M. McGlaun, S. L. Thompson, and M. G. Elrick, "CTH: A three-dimensional shock wave physics code," *Int. J. Impact Engng.*, **10**, 351-360, 1990.

[7]K. A. Dannemann, A. E. Nicholls, C. E. Anderson, Jr., S. Chocron, and J. D. Walker, "Response and characterization of confined borosilicate glass: intact and damaged," *30th Int. Conf.& Exp. on Advanced Ceramics and Composites*, The American Ceramics Society, Cocoa Beach, FL, 22-27 January, 2006.

[8]S. Chocron, J. D. Walker, A. E. Nicholls, C. E. Anderson, Jr., and K. A. Dannemann, "Constitutive model for damaged borosilicate glass," *30th Int. Conf.& Exp. on Advanced Ceramics and Composites*, The American Ceramics Society, Cocoa Beach, FL, 22-27 January, 2006.

# RESPONSE AND CHARACTERIZATION OF CONFINED BOROSILICATE GLASS: INTACT AND DAMAGED

Kathryn A. Dannemann, Arthur E. Nicholls, Charles E. Anderson, Jr., Sidney Chocron, James D. Walker
Southwest Research Institute
P.O. Drawer 28510
San Antonio, Texas 78228-0510

## ABSTRACT

The objective of this work is to determine the fundamental compression response of borosilicate glass and obtain an improved understanding of the transition from intact to damaged material to aid in glass modeling efforts. Compression experiments were conducted on borosilicate glass under confinement. An experimental technique developed for ceramics[1] is applied to glass specimens to obtain comminuted glass material by *in-situ* failure of intact or pre-damaged (by thermal shock) samples. Emphasis is on development of the experimental technique and application of additional diagnostics to characterize the response of comminuted glass. The glass samples (intact or pre-damaged) were inserted in a high-strength steel confining sleeve, and then loaded and re-loaded at quasistatic strain rates to fail the material *in-situ*. Multiple load/reload cycles were applied at successively increasing compressive loads. Strain gages mounted on the outer diameter of the confinement sleeve were used to measure hoop strain. Differences in the response of intact vs. pre-damaged glass material are evaluated and presented. Interpretation of the results within a constitutive model for borosilicate glass is presented in a companion paper by Chocron, et al.[2]

## INTRODUCTION

Existing models [3,4] for evaluating the performance of ceramics under ballistic impact do not adequately address the transition from intact to damaged material behavior owing to the lack of experimental data in this regime. Significant progress[5,6] has been made, though additional work is warranted based on the critical needs in this area. The focus of the present work is to obtain a better understanding of this transition in borosilicate glass to aid in modeling efforts.

The compressive response of various intact and powder ceramic materials (e.g., $Al_2O_3$, $AlN$, $B_4C$) has been characterized previously by numerous investigators using confinement techniques.[7,8,9,10,11] The authors recently evaluated the compressive response of intact and powder forms of SiC-N, as well as *in-situ* damaged material.[1] Experimental techniques were devised to improve understanding of the transition from intact to damaged SiC-N material. Similar techniques are applied in the present work to characterize the compressive response of borosilicate glass under confinement. Emphasis is on further development of the experimental technique, accompanied by additional diagnostics for monitoring the damage response. This includes hoop strain measurement, application of an acoustic emission technique to assist in interpretation of the experimental data, as well as interrupted compression tests for detecting the onset and accumulation of damage.

The confinement experiments are more easily performed on borosilicate glass than SiC-N due to the lower compressive strength of the glass. Damage accumulation is also more readily observed owing to the transparency of the glass. Both intact and pre-damaged borosilicate glass is characterized. Although pre-damaging of the glass is not necessary owing to its lower strength, application of the same technique to the glass as the SiC-N ceramic allowed validation of the test procedure. Hence, a thermal shock was applied to pre-damage the glass samples.

MATERIALS

The glass material evaluated was Borofloat® 33, a borosilicate glass manufactured by Schott Glass using a float process. The material was obtained from Swift Glass, Elmira, NY. The Borofloat (BF) glass evaluated consists primarily of $SiO_2$ (80.5 w/o) and $B_2O_3$ (12.7 w/o) with $Al_2O_3$ (2.5 w/o), $Na_2O$ (3.5 w/o) and $K_2O$ (0.64 w/o) and minor amounts of other oxides. These values are based on X-ray fluorescence analysis on select test samples. A low $Fe_2O_3$ content contributes to the transparency of this glass. Table 1 is a summary of some properties of the Borofloat (BF) glass.[12] These were measured on the BF glass material obtained for testing using ultrasonic velocity measurements in accordance with ASTM E494.[13] The moduli and Poisson's ratio are comparable to the values measured in the experiments discussed below.

Table 1. Borofloat® 33 Properties based on Ultrasonic Velocity Measurements[12]

| Density | 2.23 g/cc |
|---|---|
| E | 62 GPa |
| G | 26 GPa |
| Poisson's Ratio | 0.2 |
| $c_L$ | 5600 m/s |
| $c_S$ | 3400 m/s |

The Borofloat glass was provided in the form of cylindrical test samples, measuring 12.7-mm long by 6.35-mm diameter (L/D = 2). These were ground from plate material and polished to a high end optical finish to minimize surface effects owing to the inherent flaw sensitivity of glass. Flatness and parallelism of the sample ends, especially critical when testing brittle materials, was maintained to within 0.0051-mm.

Tapered WC loading anvils were also machined and obtained for tests with confining sleeves. The WC material grade used was selected for its high impact and shock resistance, and is similar to that used previously for autofrettage testing[14] of ceramics at SwRI. Annular steel confining sleeves were fabricated from maraging steel, Vascomax C350 to minimize yielding.

EXPERIMENTAL PROCEDURE

Compression tests were conducted on Borofloat glass. Most experiments were performed on confined test samples, for both intact and pre-damaged glass, to provide data on glass damage as a function of confining pressure. Several tests were also conducted on intact borosilicate glass without confinement. Comminuted material was obtained by either (i) load/reloading of pre-damaged (i.e., thermally shocked) samples under confinement or (ii) load/reloading of intact samples under confinement. All tests were conducted at quasistatic strain rates (~$10^{-3}$ to $10^{-4}$ s$^{-1}$) using an MTS servohydraulic machine. The quasistatic tests are a precursor to similar tests planned at high strain rates (~$10^3$ s$^{-1}$) using a split Hopkinson pressure bar (SHPB) technique.

Confinement tests were performed using a high-strength steel confining ring (22.25-mm long, 3.22-mm wall thickness). The inner diameter of each sleeve was honed to ensure an appropriate fit between the confining sleeve and each individual sample. The clearance between the sample and the confining sleeve was 0.0127-mm nominal. Tapered WC loading anvils were positioned at the specimen ends, as shown in Figure 1. Axial stresses and strains were monitored during testing. Axial strains were measured with an extensometer. Hoop strains were measured using strain gages attached to the outer diameter of the steel confining sleeve. Two strain gages were utilized; these were located on opposite sides of the sleeve diameter.

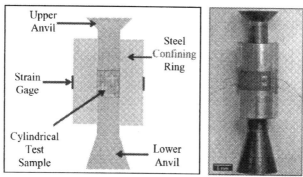

**Figure 1. Schematic on the left shows the test assembly used for confinement testing of intact and pre-damaged Borofloat glass samples. The photo on the right shows the actual test assembly with strain gages on the confining sleeve for measurement of hoop strain.**

Although pre-damaging of the glass samples was not necessary to decrease the strength of the glass material prior to confinement testing, a thermal shock procedure was utilized to validate the test procedure used to test pre-damaged SiC-N material in earlier testing[1]. Individual samples were exposed for two 0.3-h cycles at 500°C in a resistance tube furnace. Each thermal exposure was followed by an ice water quench. The transparency of the Borofloat glass allowed easier viewing of the damage due to thermal shock. Visual observation of the test samples indicated that the thermal shock technique was successful in pre-damaging the glass. Stereomicroscopy evaluations of the thermally-shocked samples prior to confinement testing revealed a crack pattern on the sample ends and throughout each test sample. A consistent damage pattern was observed from sample to sample. A representative Borofloat sample (BF-14) is illustrated in Figure 2 following the thermal shock procedure and prior to testing. The thermal shock procedure produced a crack pattern without causing a loss of sample integrity (i.e., the samples remained intact and could be readily tested). No volume change was measured between the intact and pre-damaged test samples prior to testing. The average diametral change measured on pre-damaged samples following thermal shock was 0.002-mm, resulting in volumetric changes of only 0.677-mm$^3$ (i.e., approximately 0.17% of initial volume).

Load/reload compression experiments were performed on both pre-damaged (i.e., thermally shocked) and intact BF glass samples under confinement (i.e., samples were positioned in the steel confining sleeve as shown in Figure 1). For each test sample, multiple load/reload cycles were conducted at successively increasing loads. This resulted in "comminution" of the Borofloat glass samples. All tests were conducted in displacement control at quasistatic strain rates using a compression load/release/reload sequence. The number and length of the load/reload cycles varied with each test and test sample condition. Loading increments were generally small to aid in interpretation of the test data and, as discussed in subsequent sections, allow samples for post-test evaluation. Some interrupted tests were also performed on pre-damaged BF samples to detect the onset and extent of further damage. This involved several load/reload cycles at successively increasing loads, followed by sectioning of the confining sleeve and microscopic evaluation of the test sample.

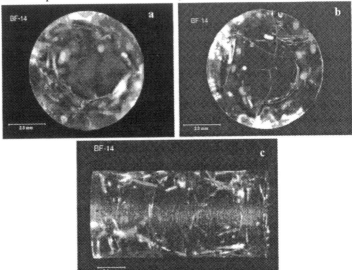

Figure 2. A thermal shock procedure was used to pre-damage the Borofloat test samples prior to load/reload testing. (a) and (b) are views of both polished ends of the sample following the thermal shock procedure and prior to confinement testing; (c) is a side view of the thermally shocked sample.

An acoustic emission (AE) technique was employed for some of the load/reload tests to assist in interpretation of the experimental data and aid in monitoring the onset and accumulation of damage. The acoustic emission system utilized is manufactured by Physical Acoustics Corporation (Princeton, NJ). The model is the MISTRAS 2001 operating system with AEDSP-32/16B digital data acquisition boards that operate at 10 MHz. The overall system can resolve and analyze acoustic events as short as 250 ns.

RESULTS
Intact
        Initial confined experiments were conducted on intact BF glass. Sample surfaces and ends were well-polished to minimize the effects of surface flaws. Multiple load/reload cycles were applied to each intact sample tested. The number of cycles accomplished varied with test sample and ranged from 5 to 10 cycles. The maximum stress attained also varied for each intact sample tested; the maximum stress applied approached 3500 MPa (i.e., the limit of the WC platens). For the confined tests conducted to date on intact samples, the test remained elastic for stresses less than 3000 MPa. Consecutive loading curves tracked closely and were quite consistent; minimal hysteresis was observed. There was minimal change in slope for the axial stress-strain curves after the initial load cycle. The elastic moduli measured from the curves were similar; the values were slightly less than determined from ultrasonic velocity measurements, reported in Table 1. Representative results (Sample BF-3) are shown in Figure 3 where five load/reload cycles are represented. The loading cycles were applied in stress

increments of approximately 500 MPa. Axial stress is plotted versus both axial strain and hoop strain. The hoop strain showed a gradual increase with stress. The test sample represented in Figure 3 did not fail. Additional testing is planned for further correlation of the damage response for intact confined samples.

To determine if bulking of the confined samples occurred with loading, volumetric changes were determined for each confined sample following testing. The volume change was determined based on calculations of the radial displacement of the steel confining sleeve. The volume decreased with increasing stress. The percent decrease in volume is shown in Figure 4 for the results shown in Figure 3. The volume decrease exceeded 4% after the $5^{th}$ loading cycle for sample BF-3. A permanent decrease of approximately 0.2% was determined after final load removal.

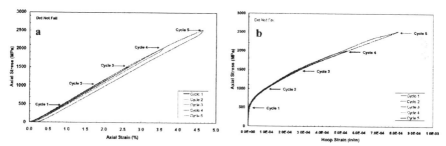

Figure 3. Representative axial stress versus strain curves (Sample BF-3) for intact Borofloat glass during quasistatic compression load/reload cycling. Note the effect of five load/reload cycles on: (a) axial strain and (b) hoop strain.

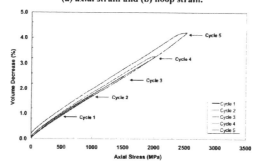

Figure 4. Volume decrease (in percent) plotted vs. axial stress for the intact sample shown in Figure 3 due to load/reload cycling.

Pre-Damaged

Confinement experiments were also conducted on pre-damaged samples using compressive load/reload cycling at quasistatic strain rates. The surface finish of these samples was less critical than for the intact samples since they were pre-damaged by thermal shock. Figure 2 is representative of the extent of damage created due to the thermal shock procedure performed prior to testing. Multiple compressive load/reload cycles were also applied to the pre-damaged test samples. The maximum applied stress approached approximately 3500 MPa for these tests. An appreciable number of load/reload cycles was accomplished, ranging from 6 to 9 cycles. The maximum axial strains achieved were in the vicinity of 7 to 8%.

The stress-strain results are quite consistent for the confinement tests conducted to date on pre-damaged samples. The response of the pre-damaged BF glass for the first few loading cycles (applied at approximate stress increments of 500 MPa) is elastic. A deviation from elastic behavior is generally observed at approximately 1800 MPa, and appears to correlate to a change in damage mechanism for the thermally shocked samples. Significant jumps in the measured hoop strains were observed in the stress-strain curves at stresses exceeding this value. Additional testing, included interrupted tests, is underway to determine the reproducibility of this stress value and whether it correlates to a significant damage event.

The results for a representative test are shown in Figure 5 where the effect of load cycling on the axial and hoop strains is evident. Pre-test photographs of this sample (BF-14) are shown in Figure 2; acoustic emission results are also discussed subsequently for this same sample. Nine compressive load/reload cycles were applied. An increase in slope is evident following the initial loading cycle. The four subsequent loading cycles demonstrate some further, but slight, decreases in slope. A significant change occurs during the fifth loading cycle. This corresponds to a large increase in hoop and axial strain, indicative of a significant damage event. The hoop strain increased almost threefold, as shown in Figure 5(b). Further load cycling (cycles 6 through 9) caused an additional decrease in slope and increases in hoop strain. These changes are indicative of further damage to the sample due to load cycling.

This was confirmed upon determination of the volume change for this pre-damaged sample. A volume decrease was also calculated for this confined test sample. A continuous volume decrease was determined for each subsequent loading cycle. Hence, bulking does not likely occur. The results are shown in Figure 6 for sample BF-14. The volume decrease was approximately 4% at a stress level of 2500 MPa. This is similar to the volumetric changes observed for the intact confined sample. The volume of the pre-damaged samples further decreased during the final loading cycle and approached a 6% decrease at maximum axial stress. A permanent volume decrease of ~1.5% was determined for sample BF-14. Similar volume decreases were observed for the other pre-damaged samples tested in confinement.

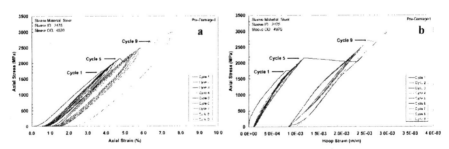

Figure 5. Axial stress versus strain curves for pre-damaged (by thermal shock) Borofloat glass (Sample BF-14) during quasistatic compression load/reload cycling. Note the effect of nine load/reload cycles on: (a) axial strain and (b) hoop strain.

The pre-damaged samples were evaluated following confinement testing with load cycling to determine the extent of damage incurred. Samples were not readily removed from the confining sleeve; separation of the sleeve was necessary for sample removal. Observation of the samples revealed significant damage, though some larger sample fragments remained. These remnants easily broke into numerous pieces with application of minimal force. The post-test condition of Sample BF-14 is illustrated in Figure 7. These photos are representative of the damage observed in other samples evaluated. The results indicate that comminuted glass material was created by *in-situ* failure of thermally shocked, pre-damaged samples.

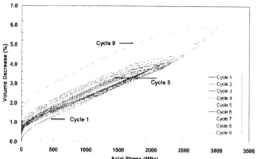

Figure 6. Volume decrease (in percent) plotted vs. axial stress for the pre-damaged sample (BF-14) shown in Figure 5 due to load/reload cycling.

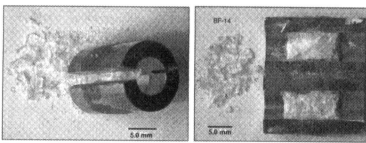

Figure 7. Post-test condition of the pre-damaged test sample (BF-14) described in Figure 5 and Figure 6. Removal of the confining sleeve (left) shows sample remnants with some test pieces remaining in the confining sleeve (right).

Microscopic evaluation of pre-damaged samples following interrupted tests with load/reload cycling at successively increasing loads was beneficial in understanding the damage response. The onset and extent of damage were detected upon inspection of the sample ends, as well as the sample length. The entire length of each sample was visible following removal of a small section of the confining sleeve. Both ends of each sample were viewed and photographed before removal of the confining sleeve. An increase in the extent of damage with increasing number of load/reload cycles (and peak stress) is evident in the photos of the sample ends in Figure 8. The photos are sequenced in increments of 500 MPa. Additional damage, beyond the pre-damage condition, was detected following an initial load/reload cycle to 500 MPa. The extent of damage increased with further cycling and loading.

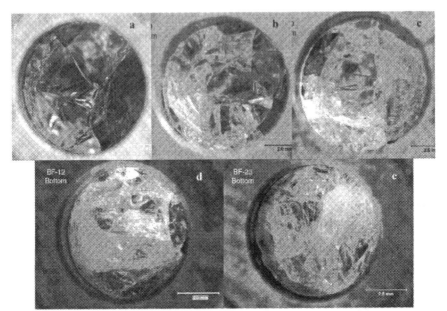

**Figure 8. End views of pre-damaged samples following interrupted load/reload tests show the extent of damage increases with stress and number of load/reload cycles. Each load/reload cycle was applied in 500 MPa increments. The maximum stress achieved for these samples is: (a) 500 MPa, (b) 1000 MPa, (c) 1500 MPa, (d) 2000 MPa, (e) 2500 MPa.**

Comparison – Intact vs. Pre-Damaged

Findings from the confinement tests performed to date on the intact and pre-damaged samples demonstrate that the load cycling technique is effective in "failing" the glass *in-situ* and creating comminuted material. Scanning electron microscopy of some of the sample remnants revealed micron-sized particles with an increase in the number of smaller particles with increasing load. This will be discussed further in a forthcoming publication.

Several unconfined compression tests were initially conducted to serve as a baseline for comparison with the confined experiments. These test results exhibited significant variation owing to the flaw sensitivity of the glass. Nevertheless, the maximum compressive strength measured for the unconfined glass samples is approximately 2000 MPa. Maximum stress levels achieved for BF glass in the confined tests approach 3500 MPa for the pre-damaged samples, and approximately 4000 MPa for the intact samples. Hence, confinement significantly increases the strength of the borosilicate glass relative to unconfined material. Load cycling of confined test samples caused a decrease in slope with a greater decline for the pre-damaged samples when compared over a similar stress range. This is illustrated in Figure 9 for an intact unconfined test sample vs. intact confined and pre-damaged confined samples. The experimental results have been applied to obtain a constitutive model for intact and damaged borosilicate glass. Derivation of the model and further discussion of the experimental results is presented in a companion paper by Chocron, et al.[2]

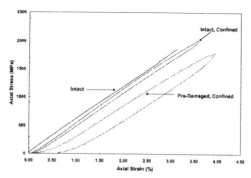

**Figure 9. Axial stress vs. axial strain comparison for unconfined intact (black) vs. load/reloaded confined samples. The curves for the confined samples are individual cycles for intact confined and pre-damaged confined samples. Note the slight decrease in slope for the pre-damaged, confined sample.**

Acoustic Emission Results – Pre-Damaged Samples

An AE technique was applied for some confinement tests to assist in data interpretation and aid in monitoring the damage response. Promising results to date have been obtained for pre-damaged confined samples during load/reload testing. Damage events appear to be detectable with the AE technique. Highlights of the recent findings are presented here. AE data were obtained and are shown for cycles 6 through 9 for Sample BF-14. Proper filtering parameters have now been determined and the technique will be more readily applied in future experiments.

The AE data obtained is summarized in Figure 10 where cumulative counts, energy, and average frequency are plotted versus time. Note the frequency data correspond to the y-scale on the right side of the plot. Correlation of the data with the loading/unloading cycles is more readily observed by overlaying the AE data on strain vs. time and stress vs. time plots. These plots are shown in Figure 11 and Figure 12, for strain and stress respectively. Cumulative counts and energy are cross-plotted vs. time in Figure 11 for the axial and hoop strains measured during loading and unloading in cycles 6 through 9. A similar plot is shown for axial stress in Figure 12(a); the frequency values are included in Figure 12 (b). Note the increase in counts, energy and frequency with loading and increased stress. Higher energy levels were recorded on the loading cycles; the energy values decreased with unloading. This is most apparent for the strain vs. time plots in Figure 11. The AE data are quite consistent until the final loading cycle when there is a significant change in hoop strain. The results correlate with significant changes in strain/stress and likely correspond to damage events. This will be confirmed with interrupted tests for sample removal and observation.

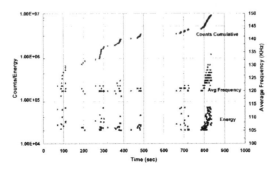

Figure 10. Acoustic emission data for cycles 6 through 9 for pre-damaged Sample BF-14, shown in Figure 5. The left axis corresponds to counts and energy; the right axis is scaled for average frequency (kHz).

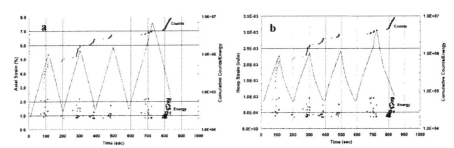

Figure 11. Acoustic emission (AE) data for load/reload cycles 6-9 for pre-damaged sample BF-14 shown in Figure 5. Counts and energy are cross-plotted with strain vs. time for (a) axial strain, and (b) hoop strain.

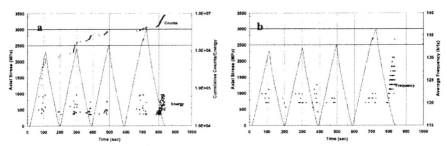

Figure 12. Acoustic emission data for load/reload cycles 6-9 for pre-damaged sample BF-14 shown in Figure 5. Axial stress vs. time curves are cross-plotted with (a) counts and energy, and (b) average frequency.

CONCLUSIONS

An innovative experimental technique is employed for conducting confinement experiments (i.e., non-ballistic tests) in compression on Borofloat glass, including intact, and pre-damaged material forms. The purpose of the experiments is to determine the fundamental compression response of intact vs. damaged glass, and to aid in deriving constants for constitutive models independent of ballistics experiments. A procedure was developed for obtaining "comminuted" material by *in-situ* failure of thermally shocked, pre-damaged samples. Comminuted material was also obtained by load/reload compression experiments on intact material. The results of load/reload experiments show a similar response for the two material types. As expected, confinement significantly increases the strength of Borofloat glass. An acoustic emission technique proved effective in detecting damage events. Additional tests are underway to better understand the damage response. This includes evaluation of samples at various stages of loading (and damage).

ACKNOWLEDGMENTS

The authors gratefully acknowledge the financial support of the US Army (Contract Number F426000-D-8037-BR02). Technical assistance and insight from Dr. Doug Templeton (RDECOM-TARDEC) is also gratefully acknowledged.

REFERENCES

1. K.A. Dannemann, A.E. Nicholls, S. Chocron, J.D. Walker, C.E. Anderson, Jr., "Compression Testing and Response of SiC-N Ceramics: Intact, Damaged and Powder", American Ceramic Society 29th International Conference on Advanced Ceramics and Composites, Cocoa Beach, FL (2005).

2. S. Chocron, K.A. Dannemann, A.E. Nicholls, J.D. Walker, C.E. Anderson, Jr., "A Constitutive Model for Damaged Glass", American Ceramic Society 30th International Conference on Advanced Ceramics and Composites, Cocoa Beach, FL (2006).

3. T. Holmquist and G.R. Johnson, "Response of Silicon Carbide to High Velocity Impact", *J. Appl. Phys.*, **91** (9), 5858-5866 (2002).

4. G.R. Johnson, T. Holmquist, "Response of Boron Carbide Subjected to Large Strains, High Strain Rates, and High Pressures", *Journal of Applied Physics*, **85** (12), 8060-8073, (1999).

5. G.R. Johnson and T.Holmquist, "Some Observations on the Strength of Failed Ceramic", American Ceramic Society 29th International Conference on Advanced Ceramics and Composites, Cocoa Beach, FL (2005).

6. S. Chocron, K.A. Dannemann, A.E. Nicholls, J.D. Walker, C.E. Anderson, "A Constitutive Model for Damaged and Powder Silicon Carbide", American Ceramic Society 29th International Conference on Advanced Ceramics and Composites, Cocoa Beach, FL (2005).

7. J. Lankford, C.E. Anderson, Jr., A.J. Nagy, J.D. Walker, A.E. Nicholls, and R.A. Page, "Inelastic Response of Confined Aluminum Oxide under Dynamic Loading Conditions", *J. Mat. Sci*, **33**, 1619-1626 (1998).

8. J. Lankford, "Compressive Strength and Microplasticity in Polycrystalline Alumina", *J. Mat. Sci.*, **12**, 791-796 (1977).

9. W. Chen and G. Ravichandran, "Static and Dynamic Compression Behavior of Aluminum Nitride under Moderate Confinement", *J. Am. Ceram. Soc.*, **79**, 579-584 (1996).

10. Y.B. Gu and G. Ravichandran, "Dynamic Behavior of Selected Ceramic Powders", *Int. J. Impact Eng.*, **32**, 1768-1785 (2006).

11. L.W. Meyer and I. Faber, "Investigation on Granular Ceramics and Ceramic Powder", *J. Phys. IV France*, **7**, Colloque C3, C3-565 – C3-570 (1997).

12. P. Patel, US Army Research Laboratory, Aberdeen, MD, personal communication, (Sept. 2005).

13. ASTM E494, "Technique for Measuring Ultrasonic Velocity in Materials", July2001.

14. J. Lankford, "Dynamic Compressive Failure of Brittle Materials under Hydrostatic Confinement", AMD-Vol. **165**, *Experimental Techniques in the Dynamics of Deformable Solids*, ASME (1993).

# CONSTITUTIVE MODEL FOR DAMAGED BOROSILICATE GLASS

Sidney Chocron, James D. Walker, Arthur E. Nicholls, Charles E. Anderson, Kathryn A. Dannemann
Southwest Research Institute
PO Drawer 28510
San Antonio, Texas 78228-0510

## ABSTRACT

An experimental technique developed in a previous paper, and consisting of testing a predamaged specimen inside a steel sleeve, is used to obtain the data to develop the constitutive equations (elastic and plastic behavior) for Borofloat® 33 glass. The glass was chosen as the specimen because it is easy to fail in that configuration. This paper first briefly summarizes the experimental technique and then shows that, if the specimen follows a Drucker-Prager plasticity model it is possible to determine, with the help of an analytical model, the elastic and plastic constants from the slopes of the axial stress vs. axial strain and axial stress vs. hoop strain curves measured in the laboratory tests. The paper determines the constants and shows how the model compares with the test data available so far. The analytical model is verified with the help of LS-DYNA in 2-D and 3-D numerical simulations. The analytical and numerical models allow a double check of different assumptions and confirm that the experimental technique is a valid procedure to determine the elastic and plastic constants. The constants can then be used in very different computations like ballistic penetration.

## INTRODUCTION

An experimental technique consisting of testing a predamaged specimen of SiC-N inside a steel sleeve allowed the determination of the constitutive elastic and plastic equations for SiC-N as shown in [1]. An analytical model was simultaneously developed [2] to help in the interpretation of the elastic part of the load. The same experimental technique is used in this work, see [3], but with predamaged glass (Borofloat® 33 ) specimens. This time, to seek a more complete interpretation of the tests, an elasto-plastic analytical model was developed together with numerical simulations using LS-DYNA in two and three dimensions.

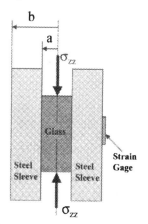

Figure 1: Experimental set-up

## EXPERIMENTAL TECHNIQUE

The experimental technique is thoroughly described in the paper by Dannemann et al. [1] so it will only be briefly summarized here for completeness. A predamaged borosilicate glass (Borofloat® 33 ) specimen with cylindrical shape is placed inside an annular steel (Vascomax) sleeve and stresses are applied at both ends of the specimen by means of an MTS machine. During the test the axial stress and strain of the specimen are being recorded. The hoop strain of the sleeve is also measured with a strain gage placed on its outer diameter and centered with the specimen, see Figure 1.

If the sleeve does not yield during the test a simple analytical model relates the hoop strain in the outer diameter with the confining pressure and, in general, the whole stress and strain state is known in both the sample and the sleeve.

There are some concerns with the experimental technique that will be addressed with the numerical and analytical models:

1) Friction between the specimen and the sleeve might artificially increase the axial strain.
2) There might be a hoop strain gradient along the axis direction in the outer diameter of the sleeve. This may cause non-uniform loading of the specimen. Since the equivalent stress is the difference of two large quantities, small errors can result in misinterpretation of the results.
3) The sleeve might be yielding at some point during the experiment, invalidating the elastic assumption.

The analytical and numerical models will also be a helpful tool to answer some interesting questions concerning the glass:

1) Does the glass bulk when it fails?
2) Do the elastic constants change from intact to predamaged glass?
3) Do the elastic constants change when cycling or further damaging the glass?
4) Does the glass fail catastrophically, or is there inelastic-plastic flow?
5) Can the inelastic-plastic flow of the glass be appropriately modeled (numerically or analytically)?
6) Can failure of the glass be modeled?

In this context the word "failure" or "catastrophic failure" means a sudden change in the elastic and/or plastic constants that describe the material.

## ELASTO-PLASTIC ANALYTICAL MODEL

An analytical model for the elastic loading was developed in a previous paper [2]. The analytical model provided a fast and easy way to estimate the elastic modulus and Poisson's ratio of the specimen during the elastic loading by means of a linear system of four equations but it could not be used when the specimen started to flow plastically.

In this paper the model has been completely rewritten in terms of increments of elastic and plastic strains. This allows the explicit calculation of the slopes of the elastic and plastic branches of the Stress vs. Strain and Stress vs. Hoop Strain curves from the elastic constants and the assumed plastic constitutive equation. In other words, the slopes measured during the tests give directly the constitutive constants of the specimen, as it will be shown.

Figure 2 shows an idealized interpretation of the test results done in [3] although only the loading branch is shown. Even though the glass tested is predamaged it is assumed in this paper that an elastic branch (1 and 1′ in Figure 2) of deformation does exist because, as shown in [3], a load-unload cycle returns to zero strain if the yield point is not attained. It is also assumed that at some point yielding starts with a small change in slope (2 and 2′), and if unload happens, a permanent deformation could be measured. During plastic flow the behavior of the material is assumed to be smooth, its elastic constants should remain the same and the yield strength is assumed to follow a Drucker-Prager model ($Y=Y_0+\beta P$, where $Y_0$ and $\beta$ are material constants). It is known that cracks initiate and grow even when stresses as small as 500 MPa are applied making it "risky" to argue that the specimens remain elastic or deform plastically. Still the authors think that, when the material is confined (as in these experiments or in ballistic penetration), even if some limited crack growth happens the specimen, in the macroscale, behaves as if it were elastic and/or plastic below a threshold where sudden jumps appear, see [3].

Tests also show sudden jumps in both hoop strain and axial strain where slopes 3 and 3′ are virtually horizontal. That part of the test is not addressed in this paper and is the subject of further research.

The analytic model is fully explained in the appendix of this paper. To make the paper less cumbersome to read only the main assumptions and results of the model will be explained in the main part of the paper.

### Assumptions

The problem is assumed to be axisymmetric and the radial displacement fields in the specimen and sleeve follow the equation $u_r(r) = Ar + B/r$, where A and B are determined by the boundary conditions for each stress $\sigma_{zz}$ and are different for the specimen and sleeve. The specimen is assumed to smoothly slip in the sleeve so no friction is considered between the sleeve and the specimen (this assumption is discussed later in the paper). The constitutive equations of the specimen are Hooke's law and the Drucker-Prager model. A uniform strain is assumed in the axial direction, i.e., the specimen and sleeve do not bend. The stress in the axial direction in the sleeve is assumed to be zero (which is a consequence of the no friction assumption).

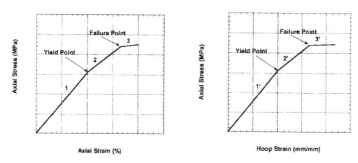

Figure 2: Idealized stress-strain curves obtained from the testing.

Given all the above assumptions it is possible to explicitly write slopes 1, 1', 2, 2' as a function of the elastic and plastic constants of specimen and sleeve:
Slope 1, which happens during elastic deformation, is given by (Eq. 29):

$$(Eq.\ 1) \qquad \left.\frac{d\sigma_z}{d\varepsilon_z}\right|_{elastic} = \frac{2\lambda d\varepsilon_r + (\lambda + 2\mu)d\varepsilon_z}{d\varepsilon_z}$$

where $\lambda$ and $\mu$ are the Lamé constants for the specimen and $d\varepsilon_z$ is the applied (known) strain in the specimen. Using (Eq. 30):

$$(Eq.\ 2) \qquad slope\ 1 = \left.\frac{d\sigma_z}{d\varepsilon_z}\right|_{elastic} = \frac{2\lambda^2}{C' - 2(\lambda + \mu)} + (\lambda + 2\mu)$$

$C'$ is a constant that relates the radial stress (or confinement pressure) in the specimen with the radial displacement of the inner part of the sleeve. $C'$ only depends on the elastic constants of the sleeve and its geometry and is explicitly given by (Eq. 25) in the Appendix.
Similarly, slope 2 is given by (Eq. 40) in the appendix:

$$(Eq.\ 3) \qquad slope\ 2 = \left.\frac{d\sigma_z}{d\varepsilon_z}\right|_{plastic} = \beta'C'\left(\gamma\left(\frac{1}{2} - \delta\right) - \frac{1}{2}\right)$$

where $\beta'$, $\gamma$ and $\delta$ are given, respectively, in (Eq. 33), (Eq. 34) and (Eq. 35).
Combining (Eq. 32) with (Eq. 30) it is possible to write slope 1':

$$(Eq.\ 4) \qquad slope\ 1' = \left.\frac{d\sigma_z}{d\varepsilon_\theta}\right|_{\substack{elastic \\ r=b}} = \frac{2\lambda^2 + (\lambda + 2\mu)(C' - 2(\lambda + \mu))}{\lambda\left(A' + \dfrac{B'}{b^2}\right)}$$

A´ and B´ are constants that depend on the elastic constants of the sleeve, $\lambda´$ and $\mu´$, and the geometry of the sleeve, see (Eq. 22) and (Eq. 23). For the plastic part of the stress vs. hoop-strain curve:

(Eq. 5)
$$\text{slope } 2´ = \frac{d\sigma_r}{d\varepsilon_\theta}\bigg|_{\substack{\text{plastic} \\ r=b}} = \frac{\beta´\, C´}{\left( A´ + \dfrac{B´}{b^2} \right)}$$

Once the slopes are measured from the experiments (Eq. 2) through (Eq. 5) provide four equations with three unknowns, the elastic constants of the specimen, $\lambda$, $\mu$ and the plastic constant of the specimen, $\beta$. The strength at zero pressure $Y_0$ does not affect the slopes but determines the onset of plastic flow and which is how it will be determined.

## NUMERICAL SIMULATIONS WITH LS-DYNA

Verification of the model is usually done in the SwRI® Engineering Dynamics Department with the "triad" approach. If tests, analytical model and numerical simulations give the same results the analytical model is considered verified, and its assumptions valid. Numerical simulations in 2-D and 3-D were performed with LS-DYNA®, a finite element code developed by LSTC. The 2-D computation did not allow a Drucker-Prager strength model so only the elastic part was implemented. In the 3-D simulations there were 14 brick elements across the radius. The grid was generated with Truegrid®. The material properties used are summarized in Table 1. The penalty factor in the "contact" card was increased to 30 to avoid any interpenetration of the materials and to be able to see some measurable hoop strain in the outer part of the sleeve right from the start of the load.

Table 1: Material properties used in the numerical and analytical simulations

| Specimen (damaged glass) properties | | Sleeve (Vascomax) properties | |
|---|---|---|---|
| Elastic Modulus (GPa) | 59 | Elastic Modulus (GPa) | 205 |
| Poisson's ratio | 0.19 | Poisson's ratio | 0.28 |
| $Y_0$ (MPa)O | 250 | Inner radius (mm) | 3.14 |
| $\beta$ | 1.63 | Outer radius (mm) | 6.31 |
| Length (mm) | 12.65 | Length (mm) | 22.24 |

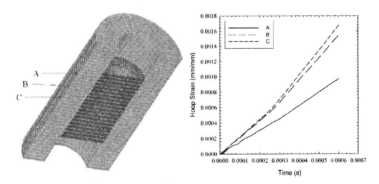

Figure 3: a) Mesh used in LS-DYNA for verification of the analytical model, b) hoop strain at three different locations in the outer surface of the sleeve

Figure 3a) shows half of the mesh used in the 3-D version of LS-DYNA. A boundary condition in displacement was applied at the top of the specimen while the motion of bottom of the specimen was constrained in the axial direction. No friction between specimen and sleeve was considered for the verification of the analytical model.

### Hoop strain gradient in the sleeve

One of the concerns of the experiment is the hoop strain gradient that happens in the outer part of the sleeve. Because the sleeve is longer than the specimen it is expected that the strains are larger in the middle vertical plane (perpendicular to the axis) of the sleeve. Figure 3b) shows strains for three tracers located on the outer surface of the sleeve, at different heights. A is located in the plane that contains the top surface of the specimen, C is located in the middle plane of the specimen and B is halfway in between. Indeed the gradient is important and care needs to be taken to place the strain gage (3.18 mm height in the axial direction and 9.6 mm long in the hoop direction) accurately in the middle plane. New tests are being prepared with a shorter sleeve to decrease the strain gradient. Also smaller strain gages will be used to avoid "averaging" through a wide area of the sleeve.

### Influence of friction between sleeve and specimen.

The friction coefficient, f, between the sleeve and the specimen was set to 0.1 and 1 in two different runs to study its influence in the interpretation of the tests. Figure 4 shows how the results change when friction is present.

For a friction coefficient f=0.1, the elastic slope of the test is only affected slightly so the Young modulus and Poisson's ratio would not change considerably. The yield point is significantly reduced (from 1.4 to 1.1 GPa), so a friction coefficient would artificially decrease the $Y_0$ measured in the test resulting in a misinterpretation of the data. The plastic slope is also affected but only in a degree similar to the scatter found during the experiments. The authors think that a friction coefficient on the order of 0.1 or less would be acceptable since the test would still give meaningful results, although it would systematically slightly increase the Poisson's ratio and decrease the slope of the Drucker-Prager curve.

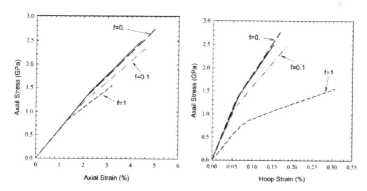

**Figure 4: Parametric study with LS-DYNA of the influence of the friction coefficient f. The red continuous line is the analytical model without friction.**

A higher friction coefficient like f=1 clearly has a tremendous impact on the tests results, see again Figure 4, invalidating the assumptions and making the results very difficult to interpret. Since the sleeve would be supporting part of the axial load transmitted from the specimen to the sleeve through friction, it should be possible to measure during the experiment an axial strain in the outer part of the sleeve. In fact the numerical simulations show that the axial stress level in the outer surface could reach around 600 MPa (or 0.1 % strain) for f=1, a value that should be easy to measure. In the future some tests with a vertical strain gage will be performed to evaluate if friction is important.

## RESULTS AND DISCUSSION

### Analysis of the experimental results using the analytical model

This part is work in progress but the results obtained so far will be presented. Four predamaged glass specimens have been tested in the confinement sleeve and analyzed by the authors. Each test consisted of multiple load-unload cycles. An example of the axial stress vs. axial strain and vs. hoop strain curves obtained during the tests are shown in Figure 5. The slopes (1,1′,2,2′), identified in Figure 2, for all the load cycles were measured. Slopes 1 and 2 in the axial stress vs. axial strain curve are simple to identify and measure. The axial stress vs. hoop strain curve starts with a bending that makes it much more difficult to identify slopes 1′ and 2′. The cause for the initial bending (a linear rise is expected) is unknown. More testing is underway with thinner and shorter sleeves to ensure that there is no contact between sleeve and anvils used to exert the axial stress and that hoop strain measurement is being done properly.

The slopes measured in each loading cycle are shown in Table 2. Slopes 2 and 2′do not always exist, because the test remained elastic, as in the first cycle. Sometimes failure hides the slopes, for example in cycles 5 and 6, so "n/a" (not available) appears in lieu of a value. The slopes were measured by fitting a linear equation to the part of the curve of interest, usually obtaining correlation coefficients of 0.999 or more.

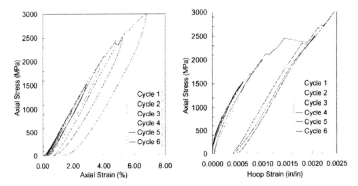

**Figure 5: Test BF-8, a) stress vs. axial strain, b) stress vs. hoop strain.**

The slopes, shown in Table 2, can be used as input of the analytical model to find E,ν and β. Depending on the cycle and test used (or by using the average) the modulus was found to range from 55 to 63 GPa, the Poisson's ratio from 0.12 to 0.19, and β from 1.6 to 2.0. The table also presents the material constants (Lamé constants, λ, μ, Slope of Drucker-Prager, β, Young Modulus, E, Shear Modulus, G and Poisson's ratio, ν) calculated from the model using each cycle slope. The properties of intact unconfined glass measured with ultrasonic and MTS procedures are also included in Table 2 for comparison.

**Table 2: Slopes of the elastic and plastic branches for test BF-8. Material constants inferred from the analytical model are also shown.**

| | Slope 1 (GPa) | Slope 2 (GPa) | Slope 1′ (TPa) | Slope 2′ (TPa) | Material constants calculated with model | | | | | |
|---|---|---|---|---|---|---|---|---|---|---|
| | | | | | λ (GPa) | μ (GPa) | β | E (GPa) | G (GPa) | ν |
| Cycle 1 | 55.84 | n/a | n/a | n/a | - | - | - | - | - | - |
| Cycle 2 | 61.01 | 56.37 | 2.38 | 2.00 | 12.00 | 25.35 | 2.01 | 58.84 | 25.35 | 0.16 |
| Cycle 3 | 61.58 | 53.88 | 1.90 | 1.59 | 13.72 | 25.02 | 1.82 | 58.90 | 25.02 | 0.18 |
| Cycle 4 | 61.96 | 49.33 | 1.69 | 1.33 | 12.98 | 25.46 | 1.66 | 59.52 | 25.46 | 0.17 |
| Cycle 5 | 60.56 | n/a | 1.62 | n/a | - | - | - | - | - | - |
| Cycle 6 | 60.35 | n/a | 1.64 | n/a | - | - | - | - | - | - |
| Average | 61.09 | 53.19 | 1.85 | 1.64 | 12.71 | 25.13 | 1.85 | 58.70 | 25.13 | 0.17 |
| Intact glass, ultrasonic measurement in the laboratory [4] | | | | | | | | 62.0-62.5 | 25.8-26.1 | .194-.207 |
| Intact glass, MTS measurement in the laboratory | | | | | | | | 57.7-62.3 | - | 0.16-0.19 |

To better assess the sensitivity of the experiment, Figure 6 presents the material constants inferred from tests BF-8, BF-11, BF-14, and BF-16 in a graphic form. Each symbol represents the values inferred from one cycle, so the plot is actually showing how the modulus and Poisson's ratio evolve during the test. The scatter in both elastic and shear modulus is reasonable with a maximum of around ±10%. The Poisson's ratio is more sensitive and has a larger scatter. Tests BF-8, 11 and 16 were similar in their procedure, increasing the final stress at each cycle. Test BF-14 was special in the sense that all the cycles went up to 2.5 GPa. The authors opinion is that there is no clear trend up or down for the moduli or Poisson's ratio. The scatter seems random supporting the fact that the elastic constants seem to stay the same, but more tests are needed to reach a final conclusion.

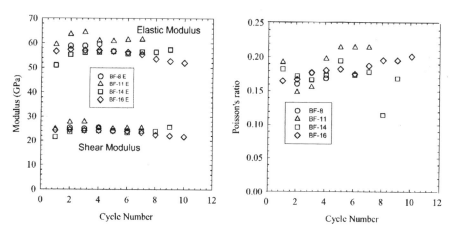

**Figure 6: a) Young Modulus and Shear Modulus evolution for each cycle inferred (one symbol per cycle) from the analytical model, four different tests (BF-8, 11, 14 and 16) shown, b) same for the Poisson's ratio.**

*Verification of the analytical model*

Figure 7 compares the results of the analytical and numerical model showing that both give very similar results. The slight differences are thought to be of numerical origin. It is very difficult to have a perfect contact between the specimen and the sleeve because, for example, the mesh of the specimen is not perfectly circular and has "corners". Nevertheless both elastic and plastic slope match very well and the authors think that the analytical model has been verified.

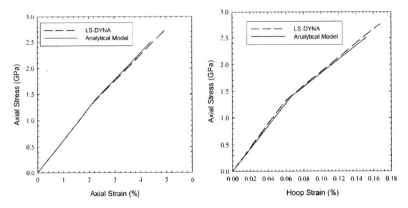

Figure 7: Verification of the analytical model by comparison with LS-DYNA results. a) Axial stress vs. axial strain, b) axial stress vs. hoop strain.

*Does the sleeve yield during the test?*

The equivalent stress is maximum in the inner part of the sleeve. Hence yielding will initiate in the inner radius of the sleeve. According to the analytical model, for the geometry presented in Table 1, the sleeve would start to yield for a hoop strain of 0.3 %. This corresponds to a confinement pressure of around 800 MPa and a pressure in the specimen of around 1.7 GPa (remember the specimen has the added component of the axial stress which the sleeve does not). Also, if the sleeve yields, the axial stress measured by the MTS should go flat. The model and the postmortem analysis confirm that the sleeve is not yielding during the test if hoop strains are kept below 0.3%.

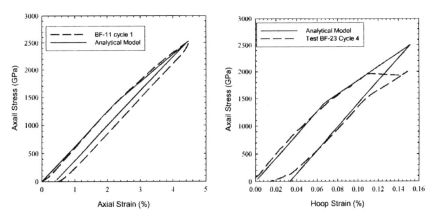

Figure 8: Comparison of analytical and experimental results for test BF-11. a) Axial stress vs. axial strain curve, only the first cycle is shown for clarity. b) Axial stress vs. hoop strain

*Comparison of the analytical elasto-plastic model with the experimental results*

Variability in the experimental results makes the determination of a set of constants a difficult task that should probably take a statistical approach. But the number of experiments is still not large enough to make a statistical determination of the constants. The following results can be considered an exercise to at least know an approximated value for $\beta$ and $Y_0$. The exercise selected consists of loading the sample up to 2.5 GPa to make it go through the elastic and plastic branches and then unload. Similar branches were selected from the experimental results to compare with the analytical model. Figure 8 compares the model and the test and cycle selected. Although it is relatively straightforward to match the axial stress vs. strain load curve, the hoop strain is much more difficult because of the initial non-linearity and the steps (interpreted as failures) that happen. Nevertheless, the match with the test is good indicating that the material constants used for the simulations are at least realistic.

*Volume change when the specimen fails*

The volume of the specimen is known at any time during the test with the help of the strain gage placed in the sleeve and (Eq. 32) and (Eq. 15), which only involve the elastic sleeve. The change of volume $\Delta V$ measured in the test can be compared with the change of volume predicted by the model, which is of elastic origin. If both $\Delta V$ were the same, then the volume change is due to elastic compression and failure does not affect the volume.

Figure 9 compares $\Delta V$ for the model and test BF-11. The first cycle of the test is not shown because it reached very low stress. The following two cycles are shown but are shifted 0.7% to compensate for the non-linear initial part. Clearly $\Delta V$ is very similar for both, leading to the conclusion that, since the postmortem analysis of the specimen shows failure, failure does not change the volume.

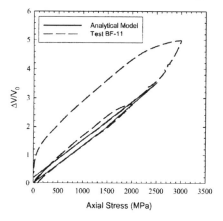

**Figure 9 Comparison of the change of volume $\Delta V$ calculated with the model and $\Delta V$ measured during the test**

*Constitutive model*

The ultimate objective of the project is to find a constitutive equation for damaged glass that could be used in numerical models to predict ballistic results. The results are still preliminary but the elastic constants range found in this paper is: $E = 54\text{-}62$ GPa, and $v = 0.16\text{-}0.2$, meaning that the elastic constants of damage glass are very similar to the intact constants. The plastic part (Drucker Prager) constants range is: $Y_0 = 250$ MPa, and $\beta = 1.6 - 2.0$. It is expected that the constant range will vary a little during the ongoing research and when more experimental results (for example with thinner and shorter sleeves to be used) become available.

Constitutive Model for Damaged Borosilicate Glass

## SUMMARY AND CONCLUSIONS

This work has shown that testing predamaged specimens in a confined sleeve is a sound experimental technique to determine the constitutive equation (elastic and plastic parts) of the specimen. With the help of the analytical and numerical models developed it is straightforward to interpret the results of the experiments and show that the assumptions are realistic and the constants obtained useful.

## ACKNOWLEDGMENT

The authors would like to acknowledge Doug Templeton from TARDEC for his support and helpful discussions.

## APPENDIX TO THE ELASTO-PLASTIC ANALYTICAL MODEL

The assumed Drucker-Prager constitutive equation during plastic deformation is of the form:

(Eq. 6)
$$Y = Y_0 + \beta P$$

where Y is the yield strength, $Y_0$ the tensile strength at zero pressure, $\beta$ the slope and P the pressure in the specimen.

Since there is cylindrical symmetry radial and hoop stresses are the same ($\sigma_r = \sigma_\theta$) so the equivalent stress in the specimen can be written, assuming shear stresses are zero:

(Eq. 7)
$$\sigma_{eq} = \sqrt{3J_2} = |\sigma_r - \sigma_z|$$

and the yield condition in incremental form is:

(Eq. 8)
$$d\sigma_r - d\sigma_z = \beta\left(-\frac{1}{3}d\sigma_z - \frac{2}{3}d\sigma_r\right)$$

Stresses are negative in compression so since $\sigma_z$ is expected to be bigger than $\sigma_r$ in absolute value, $\sigma_r$-$\sigma_z$ is expected to be positive. From Hooke's law the incremental stresses can be written:

(Eq. 9)
$$(\lambda + 2\mu)\ d\varepsilon_z^e + 2\lambda d\varepsilon_r^e = d\sigma_z$$

(Eq. 10)
$$2(\lambda + \mu)\ d\varepsilon_r^e + \lambda d\varepsilon_z^e = d\sigma_r$$

where again it is assumed that $d\varepsilon_r = d\varepsilon_\theta$ and $d\varepsilon^e$ denotes elastic strain. $\lambda$ and $\mu$ are the Lamé constants for the specimen material.

For the plastic strains in the specimen, conservation of volume yields:

(Eq. 11)
$$2\ d\varepsilon_r^p + d\varepsilon_z^p = 0$$

And the total strain is given by elastic plus plastic strain:

(Eq. 12)
$$d\varepsilon_z = d\varepsilon_z^p + d\varepsilon_z^e$$

(Eq. 13)
$$d\varepsilon_r = d\varepsilon_r^p + d\varepsilon_r^e$$

Assuming the friction is small between the specimen and the sleeve, the sleeve exerts only a radial stress on the specimen, which will be shown below to be proportional to the displacement of the inner radius of the sleeve:

(Eq. 14)
$$d\sigma_r = C'\frac{\overline{du}_r}{a}$$

where C' is a proportionality constant that only depends on the elastic constants of the sleeve and its inner and outer radius. The bar on the $u_r$ denotes that $u_r$ is the displacement of the inner radius of the sleeve, i.e. $\overline{u}_r \equiv u_r\left(r = a\right)$. Note that $\sigma_r$ is uniform in the specimen (not in the sleeve).

The last equation needed is the relation between the radial strain and the displacement of the inner radius of the sleeve that, assuming small displacements is:

(Eq. 15)
$$d\overline{u}_r = a\ d\varepsilon_r\big|_{r=a}$$

(Eq. 8) through (Eq. 15) constitute a system of eight equations with eight unknowns: $d\sigma_{rr}$, $d\sigma_{zz}$, $d\varepsilon_z^e$, $d\varepsilon_z^p$, $d\varepsilon_r^e$, $d\varepsilon_r^p$, $d\varepsilon_r$, $d\overline{u}_r$, with $d\varepsilon_z$ given applied to the end of the specimen.

*Determination of the proportionality constant C'*

In this section the objective is to find the stress in the sleeve as a function of the displacement of the inner radius of the sleeve. The displacement field assumed is of the form [2]:

(Eq. 16)
$$u_r = Ar + \frac{B}{r}$$

where A and B are constants in the sleeve, to be determined for each loading condition, and r is the distance between the axis and any point in the sleeve where the displacement is being asked. The boundary conditions for the sleeve are: 1) a given displacement in the inner radius, 2) stress free in the outer radius and 3) the stress in the axial direction in the sleeve is zero

(Eq. 17)
$$u_r(r = a) = \bar{u}_r$$

(Eq. 18)
$$\sigma_r(r = b) = 0$$

(Eq. 19)
$$\sigma_z(r) = 0$$

where the stresses are given, anywhere in the sleeve by (see [2]):

(Eq. 20)
$$\sigma_r = (\lambda' + 2\mu')\varepsilon_r + \lambda'\varepsilon_{\theta\theta} + \lambda'\varepsilon_z = (\lambda' + 2\mu')\left(A - \frac{B}{r^2}\right) + \lambda\left(A + \frac{B}{r^2}\right) + \lambda'\varepsilon_z$$

(Eq. 21)
$$\sigma_z = \lambda'\varepsilon_{rr} + \lambda'\varepsilon_{\theta\theta} + (\lambda' + 2\mu')\varepsilon_z = \lambda\left(A - \frac{B}{r^2}\right) + \lambda\left(A + \frac{B}{r^2}\right) + (\lambda' + 2\mu')\varepsilon_z$$

$\lambda'$ and $\mu'$ are the Lamé constant of the sleeve.
Applying the boundary conditions to the two last equations gives the constants A and B:

(Eq. 22)
$$A = A'\frac{\bar{u}_r}{a} \quad \text{where} \quad A' \equiv \frac{(\lambda' + 2\mu')\, a^2}{b^2(3\lambda' + 2\mu') + a^2(\lambda' + 2\mu')}$$

(Eq. 23)
$$B = B'\frac{\bar{u}_r}{a} \quad \text{where} \quad \frac{B'}{a^2} = \frac{(3\lambda' + 2\mu')\, b^2}{b^2(3\lambda' + 2\mu') + a^2(\lambda' + 2\mu')}$$

$C'$ comes from finding the radial stress in the inner diameter of the sleeve:

(Eq. 24)
$$\sigma_r(r = a) = 2\left((\lambda' + \mu')\, A' - \mu\frac{B'}{a^2}\right)\frac{\bar{u}_r}{a}$$

$C'$ is defined as:

(Eq. 25)
$$C' \equiv 2\left((\lambda' + \mu')\, A' - \mu\frac{B'}{a^2}\right)$$

so that:

(Eq. 26)
$$\sigma_r(r = a) = C'\frac{\bar{u}_r}{a} \quad \text{or} \quad d\sigma_r(r = a) = C'\frac{d\bar{u}_r}{a}$$

Again note that if r=a the radial stress in the specimen (which is constant in the specimen) is equal to that of the sleeve.
So the constants A', B', and C' are all known and functions of the elastic constants of the sleeve and its inner and outer radius. Consequently the radial stress is easily calculated with (Eq. 20). The radial and hoop strain can be calculated from:

(Eq. 27)
$$\varepsilon_r = \frac{\partial u_r}{\partial r} = \left(A' - \frac{B'}{r^2}\right)\frac{\bar{u}_r}{a} \quad \text{or} \quad d\varepsilon_r = \left(A' - \frac{B'}{r^2}\right)\frac{d\bar{u}_r}{a}$$

(Eq. 28)
$$\varepsilon_\theta = \frac{u_r}{r} = \left(A' + \frac{B'}{r^2}\right)\frac{\bar{u}_r}{a} \quad \text{or} \quad d\varepsilon_\theta = \left(A' + \frac{B'}{r^2}\right)\frac{d\bar{u}_r}{a}$$

*Explicit solution while the specimen remains elastic*

Let's find explicitly stresses and strains in the specimen and the sleeve while the specimen is elastic. From Hooke's law it is possible to write:

(Eq. 29)
$$\begin{cases} d\sigma_r = 2(\lambda + \mu)d\varepsilon_r + \lambda d\varepsilon_z \\ d\sigma_\theta = d\sigma_r \\ d\sigma_z = 2\lambda d\varepsilon_r + (\lambda + 2\mu)d\varepsilon_z \end{cases}$$

where $d\varepsilon_z$ is given.
Since $d\sigma_r = C'd\bar{u}_r/a$, which is constant in the specimen, and, from the definition of radial strains and for small displacements, $d\bar{u}_r = ad\varepsilon_r$, it is possible to find, using the first of (Eq. 29):

(Eq. 30)
$$d\varepsilon_r = \frac{\lambda}{C' - 2(\lambda + \mu)}d\varepsilon_z$$

which is constant in the specimen because the displacement field $u_r$ in the specimen is assumed to be proportional to $r$ so $\varepsilon_r(r)$=constant. (Eq. 30) allows calculating all the stresses in the specimen with (Eq. 29), for a given $d\varepsilon_z$.

An interesting quantity to calculate is the hoop strain in the outer part of the sleeve, which is measured with a strain gage during the tests. From the displacement field of the sleeve, it is possible to write, anywhere in the sleeve:

(Eq. 31)
$$d\varepsilon_\theta^{sleeve} = \left(A' + \frac{B'}{r^2}\right)\frac{d\bar{u}_r}{a} = \left(A' + \frac{B'}{r^2}\right) d\varepsilon_r\Big|_{r=a}^{specimen}$$

(Eq. 32)
$$d\varepsilon_\theta^{sleeve}\Big|_{r=b} = \left(A' + \frac{B'}{b^2}\right) d\varepsilon_r\Big|_{r=a}^{specimen}$$

*Explicit solution while the specimen is deforming plastically*

All the unknowns discussed above can be explicitly solved. First define some convenient constants:

(Eq. 33)
$$\beta' \equiv \frac{1 + \frac{2}{3}\beta}{1 - \frac{1}{3}\beta}$$

where it is recalled that $\beta$ is the slope of the yield strength as a function of pressure.

(Eq. 34)
$$\delta \equiv \frac{\lambda(\beta' - 1) - 2\mu}{2(\lambda(\beta' - 1) + \beta'\mu)}$$

where $\lambda$ and $\mu$ are the Lamé constants of the specimen.

(Eq. 35)
$$\gamma \equiv \frac{C'}{2\left(2\delta(\lambda + \mu) - \lambda + C'\left(\frac{1}{2} - \delta\right)\right)}$$

Given the above constants it is possible to explicitly write the unknowns as a function of $d\varepsilon_z$, the applied axial strain:

(Eq. 36)
$$d\varepsilon_z^e = \gamma d\varepsilon_z$$

(Eq. 37)
$$d\varepsilon_r^e = -\gamma\,\delta\,d\varepsilon_z$$

(Eq. 38)
$$d\varepsilon_r = \left(\gamma\left(\frac{1}{2} - \delta\right) - \frac{1}{2}\right)d\varepsilon_z$$

(Eq. 39)
$$d\sigma_r = C'\,d\varepsilon_r = C'\left(\gamma\left(\frac{1}{2} - \delta\right) - \frac{1}{2}\right)d\varepsilon_z$$

(Eq. 40)
$$d\sigma_z = \beta'\,d\sigma_r = \beta'C'\left(\gamma\left(\frac{1}{2} - \delta\right) - \frac{1}{2}\right)d\varepsilon_z$$

(Eq. 41)
$$d\bar{u}_r = a\,d\varepsilon_r = a\left(\gamma\left(\frac{1}{2} - \delta\right) - \frac{1}{2}\right)d\varepsilon_z$$

## REFERENCES
1. Kathryn Dannemann, S. Chocron, A. Nicholls, J. Walker, C. Anderson, Compression testing and response of SiC-N ceramics: intact, damaged and powder, Proceedings of Advanced Ceramics and Composites Conference, January 23-28, 2005, Cocoa Beach, Florida
2. Sidney Chocron, Kathryn A. Dannemann, Arthur E. Nicholls, James D. Walker, Charles E. Anderson, A Constitutive Model For Damaged And Powder Silicon Carbide, Proceedings of Advanced Ceramics and Composites Conference, January 23-28, 2005, Cocoa Beach, Florida
3. Kathryn Dannemann, A. Nicholls, C. Anderson, S. Chocron, J. Walker, Response and Characterization of Confined Borosilicate Glass: Intact and Damaged, Proceedings of Advanced Ceramics and Composites Conference, January 23-27, 2006, Cocoa Beach,
4. ARL, private communication, 2005

# REACTION SINTERED LiAlON

Raymond A. Cutler and R. Marc Flinders
Ceramatec, Inc.
2425 South 900 West
Salt Lake City, Utah 84119

## ABSTRACT

Recent results suggest that $Li^+$ can substitute for $Al^{3+}$ in the gamma-AlON structure. $LiAl_5O_8$ is isostructural with gamma-AlON above 1290°C after it goes through an ordered-disordered phase transformation and has a similar lattice parameter. A previous study showed the benefits of reaction sintering LiAlON starting with $LiAl_5O_8$ as compared to using $Li_2O$. The present study compared $Li_2O$, $LiAlO_2$, $LiAl_5O_8$, and $LiAl_{11}O_{17}$ at identical lithium levels when reaction sintering LiAlON using $Al_2O_3$ and AlN as the other reactants. X-ray diffraction was used to monitor phase changes as a function of sintering temperature. LiAlON formed by 1550°C in all lithium containing materials, with nearly full conversion by 1650°C, while AlON did not form until 1750°C under identical sintering conditions for the material without lithium. The starting source of the lithium did not control the ability to form LiAlON as was previously hypothesized. Lithia or lithium aluminate additions, however, are advantageous for reaction sintering $Al_2O_3$ and AlN since the volume expansion during sintering is avoided.

## INTRODUCTION

LiAlON was recently reported by Clay, et al.[1] as a solid solution where $Li^+$ substitutes for $Al^{3+}$ creating cation vacancies at low lithium levels. McCauley[2,3] and Corbin[4] pioneered reaction sintering work on AlON nearly thirty years ago where $Al_2O_3$ and AlN react to form a cubic spinel structure

$$\left(\frac{4-x}{3}\right)Al_2O_3 + xAlN \rightarrow Al_{\frac{(8+x)}{3}}V_{Al\left(\frac{1-x}{3}\right)}O_{4-x}N_x \tag{1}$$

creating Al cation vacancies to allow charge neutrality. $Al_2O_3$, $LiAl_5O_8$, and AlN were reaction sintered. If all of the lithia were to leave the sample during reaction sintering, then the reaction is simply given as:

$$\left(\frac{8-2x}{15}\right)LiAl_5O_8 + xAlN \rightarrow Al_{\frac{(8+x)}{3}}V_{Al\left(\frac{1-x}{3}\right)}O_{4-x}N_x + \left(\frac{4-x}{15}\right)Li_2O\uparrow \tag{2}$$

The work of Clay, et al.[1], however, indicated that all of the lithia did not leave the sample. Electroneutrality for cation vacancy formation, $V^c$, requires that if x and y range between 0 and 1 in

$$\left(Li_yAl_{(1-y)}\right)_{3-z}V_z^cO_{4-x}N_x \tag{3}$$

then $z = \dfrac{1-x-6y}{3-2y}$. An analogous requirement for anion vacancy formation was proposed and could be expected to dominate at higher levels of lithia.[1] However, the most intriguing aspect of the work of Clay, et al.[1] was the possibility that the disordered high-temperature phase of $LiAl_5O_8$, which is isostructural with $\gamma$-AlON above 1290°C,[5] could act as a nucleation site for the spinel formation when AlN and $Al_2O_3$ are reaction sintered. Since it is well understood that anion diffusion is rate controlling in the reaction sintering of AlON[6] it is of interest to see whether the presence of $LiAl_5O_8$ affects the reaction kinetics. The objective of this work was to determine if there is any advantage in using zeta alumina ($LiAl_5O_8$) when forming LiAlON, as compared to $Li_2O$ or other lithium aluminates. Reaction-sintered AlON formed using 25 mol. % AlN was therefore compared with reaction-sintered LiAlON where the AlN content remained constant at 25 mol. % while some of the $Al_2O_3$ was replaced with $Li_2CO_3$, $LiAlO_2$, or $LiAl_5O_8$ keeping the lithium content constant. The β-alumina-like composition $LiAl_{11}O_{17}$, which is reported to be stable between 1750 and 1970°C,[7] is unquenchable and therefore consists of $LiAl_5O_8$ and $Al_2O_3$ at room temperature. $LiAlO_2$ is stable to 1785°C,[7] but will react with $Al_2O_3$ to form $LiAl_5O_8$ at elevated temperature. Understanding these limitations, this work was undertaken in an effort to understand whether the enhanced optical transparency for LiAlON compared with AlON compositions made under identical conditions,[1] was merely the enhanced grain growth or was influenced by the isostructural zeta alumina at elevated temperatures.

EXPERIMENTAL PROCEDURES

$LiAlO_2$, $LiAl_5O_8$, and "$LiAl_{11}O_{17}$" were prepared by reacting appropriate amounts of $Li_2CO_3$ (Aldrich grade 25,582-3) and $Al_2O_3$ (Sasol North America grade SPA-0.5) using polyvinyl pyrrolidone (PVP) as a dispersant in isopropanol as described previously.[1] The powders were vibratory milled for 72 hours with Y-TZP media inside high density polyethylene (HDPE) containers prior to calcining the dried powder at 1000°C for 10 hours. The calcined powders were milled an additional 72 hours with surface areas of 7.8, 8.7, and 10.7 $m^2/g$ for the $LiAlO_2$, $LiAl_5O_8$, and "$LiAl_{11}O_{17}$" powders, respectively. The $LiAlO_2$ and $LiAl_5O_8$ were single-phase based on X-ray diffraction while the "$LiAl_{11}O_{17}$" powder was composed as $LiAl_5O_8$ and $Al_2O_3$, as expected.

Five compositions, as listed in Table I, were prepared at 25 mol. % AlN, with four of the materials having identical lithium concentrations and the control sample without any lithium addition. The AlN and $Al_2O_3$ powders were the same grades as used in making the $LiAlO_2$, $LiAl_5O_8$, and "$LiAl_{11}O_{17}$" additives. The powders were vibratory milled for 72 hours with 0.5 wt. % PVP dispersant in 70 grams reagent grade isopropanol in HDPE containers. The powders were dried and then slurried in hexane with 2 wt. % paraffin, based on the solids contents of the Table I compositions, before stir drying and screened through a nylon sieve. The lubricated powders were pressed uniaxially at 35 MPa followed by isostatic pressing at 250 MPa. The paraffin and dispersant were removed by heating in air to 600°C for one hour. Green dimensions after debinderization were approximately 31 mm in diameter by 5 mm in thickness, with green densities ranging between 2.2 and 2.3 g/cc for all compositions.

**Table I**
**Compositions Prepared**

| Code | Mass (g) | | | | | |
|------|-----|-----------|-------------|----------|----------|----------------|
| | AlN | $Al_2O_3$ | $Li_2CO_3$ | $LiAlO_2$ | $LiAl_5O_8$ | "$LiAl_{11}O_{17}$" |
| Control | 17.52 | 130.72 | --- | --- | --- | --- |
| $Li_2O$ | 17.52 | 130.72 | 4.11 | --- | --- | --- |
| $LiAlO_2$ | 17.52 | 125.06 | --- | 7.33 | --- | --- |
| $LiAl_5O_8$ | 17.52 | 102.48 | --- | --- | 30.00 | --- |
| $LiAl_{11}O_{17}$ | 17.52 | 68.42 | --- | --- | --- | 64.01 |

Sintering was carried out in pyrolytic BN crucibles with parts packed in powder. The powder bed was a mixture of 25 wt. % BN and 75 wt. % AlON formed by reacting alumina and aluminum nitride. The purpose of the BN was to keep the AlON from sintering together. The purpose of the packing powder was to protect the AlON from the graphite in the furnace and to reduce the tendency for the material to volatilize. Sintering temperatures ranged from 1250-2050°C with one hour isothermal holds. A slight overpressure (0.5 atm) of nitrogen was applied above 1250°C.

Density was measured using the Archimedes' method. Rietveld analysis[8,9] was used to determine phases present in the densified samples with X-ray diffraction patterns collected from 15-75° $2\Theta$ using Cu $K_\alpha$ radiation. Scanning electron microscopy was used to assess grain size and fracture mode.

## RESULTS AND DISCUSSION

Figure 1 shows the linear shrinkage, density, open porosity, and weight loss as a function of sintering temperature. In accord with the results of Clay et al.[1], the control sinters more rapidly than the sample containing zeta alumina, with densification of the $Al_2O_3$-AlN occurring prior to the transformation to $\gamma$-AlON. The primary densification of the control sample has occurred by 1550°C. The 5.0 % volume expansion expected due to Reaction (1) is clearly evident in both the shrinkage and density measurements between 1650 and 1750°C, as shown in Figure 1.

Lithia additions, regardless of the source, slow down the sintering kinetics dramatically. Sintering using the "$LiAl_{11}O_{17}$" additive is slightly faster above 1350°C than with the other lithium-containing additives. No expansion is noted when lithia is present since the slower kinetics permits conversion to the spinel structure prior to final densification. This is advantageous for making large parts, such as armor, where large volume changes can induce cracking. Weight loss measurements indicated that all of the compositions retained most of their lithia at 1850°C, as the theoretical loss was 1.1 %. When the weight loss associated with volatilization of the control was factored in, 80-90 % of the lithia remains in compositions sintered at 1850°C using $LiAlO_2$, $LiAl_5O_8$, or "$LiAl_{11}O_{17}$" additives.

Lejus and Collongues[5] showed that cubic $LiAl_5O_8$ undergoes an order-disorder phase transformation at 1290°C from the ordered low-temperature phase (space group $P4_332$ with a=7.908Å) to the disordered high-temperature phase (space group Fd3m with a=7.925Å). The eutectoid decomposition temperature reported in the literature for AlON varies with Willems et al.[10] determining it to be 1640±10°C, Hillert and Jonsson[11] calculating it to be 1627°C, Qui and Metselaar[12] lowering the value to 1612°C, and Nakao, et al.[13] measuring 1630±4°C. In any event, by monitoring differences in X-ray diffraction between 1250 and 1350°C, where the order-disorder transformation occurs, and between 1550 and 1650°C, where the $\gamma$-AlON phase becomes stable, it should be possible to tell whether zeta alumina influences the phase stability of AlON or whether the addition of lithium simply enhances grain growth, which results in enhanced transparency as discussed by Clay et al.[1]

Figure 2 shows X-ray patterns for the five materials as a function of sintering temperature. Heating to 1250°C was sufficient to allow the $Li_2CO_3$ and $LiAlO_2$ to react with the alumina so that the compositions showed similar phases, as evidenced by the Rietveld data in Figure 3. No $Li_2O$ or $LiAlO_2$ were identified by XRD, although some free $ZrO_2$, due to milling contamination and $Al(OH)_3$, likely due to reaction with water during density measurements, were

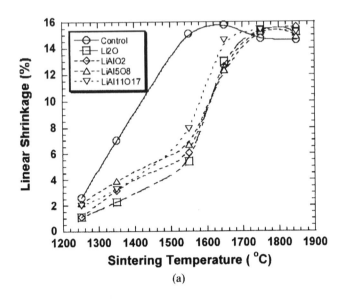

(a)

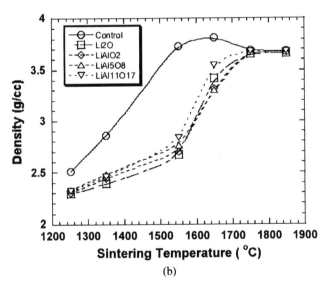

(b)

Figure 1. Linear shrinkage, density, open porosity, and weight loss as a function of sintering temperature. (a) Linear shrinkage, and (b) density.

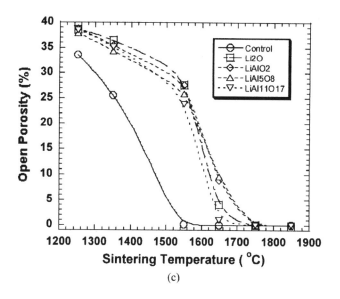

(c)

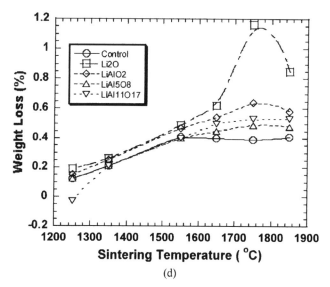

(d)

Figure 1 (continued). Open porosity (c) and weight loss (d) as a function of sintering temperature.

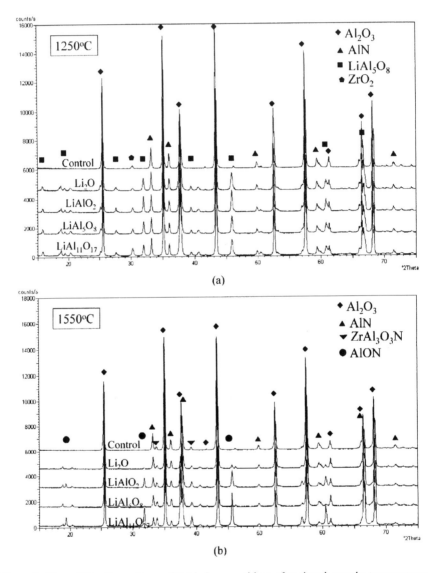

Figure 2. X-ray diffraction patterns of Table I compositions after sintering at the temperatures indicated. (a) 1250°C, and (b) 1550°C.

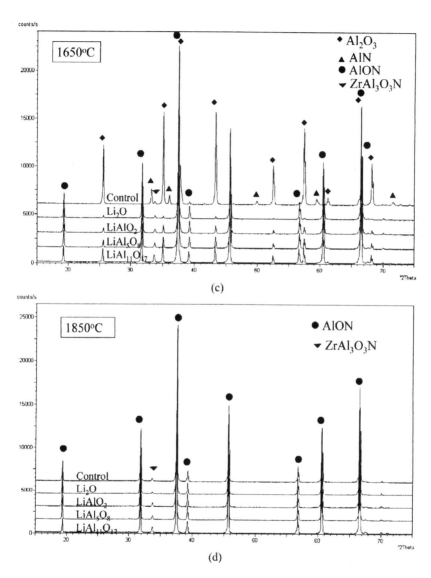

Figure 2 (continued). XRD patterns for samples sintered at (c) 1650°C or (d) 1850°C.

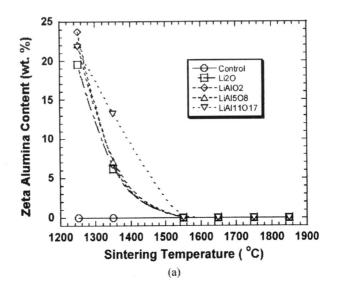

(a)

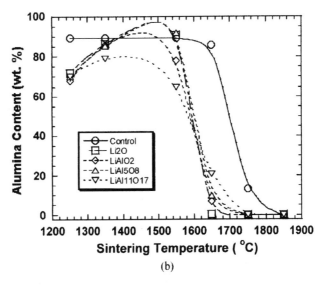

(b)

Figure 3. Rietveld fitting of XRD data showing phases present as a function of temperature. (a) LiAl$_5$O$_8$, and (b) Al$_2$O$_3$.

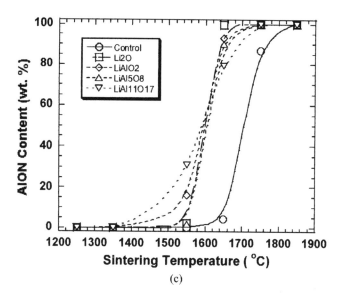

(c)

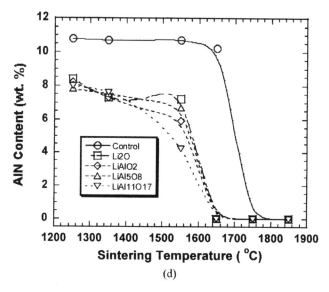

(d)

Figure 3 (continued). Rietveld fitting of AlON (c) and AlN (d) contents.

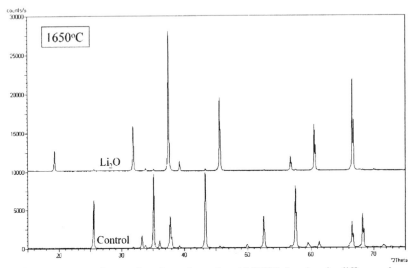

Figure 4. XRD patterns for reaction-sintered samples at 1650°C showing the difference in phases present between the two materials. The upper curve containing $Li_2O$ has little $Al_2O_3$ and consists primarily of LiAlON whereas the bottom curve still consists of $Al_2O_3$ and AlN consistent with the data shown in Figure 1.

obvious in the 1250°C pattern. This resulted in all four lithium-containing materials having similar behavior, consistent with Figure 1.

Surprisingly, the amount of zeta alumina decreased upon heating to 1350°C, as shown in Figure 3(a), at the expense of increased alumina (see Figure 3(b)). The expectation was that the disordered $LiAl_5O_8$ phase, which is stable above 1290°C, would result in additional zeta alumina, or at least the same amount, for the samples sintered at 1350°C. Zeta alumina decreased with increasing temperature and was completely gone at 1550°C. However, by 1550°C, well below the reported stability temperature for AlON, it was obvious that the lithium containing materials allowed the LiAlON to form (see Figure 2(b) and Figure 3(c)). When heating to 1650°C, where the AlON is kinetically hindered from forming in the control sample, as shown in Figures 2(c) and 3(c), it was obvious that there was a difference between LiAlON and AlON samples. This difference is illustrated in Figure 4, which compares the control sample without lithium additions with the sample made using lithium carbonate. Heating to 1750°C results in conversion to AlON or LiAlON with the exception of the $ZrO_2$, which reacts top form an oxynitride upon heating above 1350°C. JCPDS card 48-1638 ("$ZrAl_3O_3N$") gave a good fit for the zirconium-containing phase.

Fracture surfaces of the samples sintered at 1850°C are displayed in Figure 5 using backscattered imaging, which reveals the grain size. Surprisingly, the control sample had a larger grain size than the lithium-containing materials as well as showed more transgranular fracture. The difference in fracture mode may be the result of grain size or may simply be affected by the zirconia contaminates, which precipitate out at grain boundaries. It is obvious

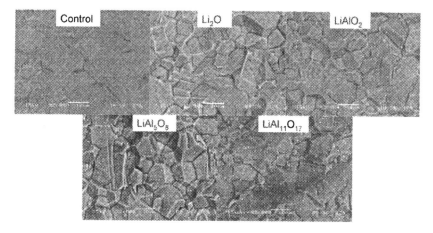

Figure 5. SEM images of fracture surfaces from the samples sintered at 1850°C showing that the grain size is larger in the control material than in the lithium containing materials.

that the milling was too aggressive and that high-purity alumina media rather than zirconia balls should have been used for the powder processing. The grain size of the control sample is larger than 10 µm whereas the lithium-containing materials all have grain size less than 10 µm. It therefore appears that the lithium can stabilize the AlON structure by going into solid solution. The starting lithium source is not important in this processing route, which results in zeta alumina as an intermediate compound in all applications. The mechanism for the stabilization is not apparent from the XRD and sintering data.

CONCLUSIONS

Reaction sintering with a variety of lithium-containing starting powders resulted in LiAlON formation below 1600°C, whereas transformation to AlON was observed at temperatures in excess of 1650°C. While zeta alumina was a precursor for all of the lithium containing materials, it was not apparent that the order-disorder transformation to the Fd3m space group was important for the $LiAl_5O_8$ as the amount of the material decreased between 1250 and 1350°C. A small amount of zeta alumina, however, may enhance the formation of LiAlON. Contrary to expectation, the lithium-containing materials were finer-grained than the material with no lithium added although zirconia contamination from the milling media may have influenced the results. By 1850°C, all materials showed similar XRD patterns. The major advantage of adding lithia or lithium aluminate to the starting composition is the ability to use a reaction sintering approach without full densification occurring prior to the volume expansion associated with the transformation.

ACKNOWLEDGEMENT

Appreciation is expressed to Lyle Miller and Angela Anderson of Ceramatec for help with X-ray diffraction work and density measurements, respectively.

REFERENCES
[1] D. Clay, D. Poslusny, M. Flinders, S. D. Jacobs, and R. A. Cutler, "Effect of $LiAl_5O_8$ Additions on the Sintering and Optical Transparency of LiAlON," *J. Eur. Ceram. Soc.*, **26**, 1351-62 (2006).

[2] J. W. McCauley, "A Simple Model for Aluminum Oxynitride Spinels," *J. Am. Ceram. Soc.*, **61**[7-8], 372-73 (1978).

[3] J. W. McCauley, and N. D. Corbin, "Phase Relations and Reaction Sintering of Transparent Cubic Aluminum Oxynitride Spinel (ALON)," *J. Am. Ceram. Soc.*, **62**[9-10], 476-79 (1979).

[4] N. D. Corbin, "Aluminum Oxynitride Spinel: A Review," *J. Eur. Ceram. Soc.*, **5**[3], 143-54 (1989).

[5] A. M. Lejus and R Collongues, "The Structure and Properties of Lithium Aluminates," *Compt. Rend.*, **254**, 2005-7 (1962).

[6] S. Bandyopadhyay, G. Rixecker, F. Aldinger, S. Pal, K. Mukherjee, and H. S. Maiti, "Effect of Reaction Parameters on $\gamma$-AlON Formation from $Al_2O_3$ and AlN," *J. Am. Ceram. Soc.*, **84**[4], 1010-12 (2002).

[7] L. P. Cook and E. R. Plante, "Phase Diagram for the System Lithia-Alumina," *Ceram. Trans.* **27**, 193-222 (1992).

[8] H. M. Rietveld, "A Profile Refinement Method in Neutron and Magnetic Structures," *J. Appl. Crystallogr.*, **2**, 65-71 (1969).

[9] D. L. Bish and S. A. Howard, "Quantitative Phase Analysis Using the Rietveld Method," *J. Appl. Crystallogr.*, **21**, 86-91 (1988).

[10] H. X. Willems, M. M. R. M. Hendrix, G. de With, G. and R. Metselaar, "Thermodynamics of AlON II: Phase Relations," *J. Mater. Sci.*, **10**, 339-46 (1992).

[11] M. Hillert and S. Jonson, "Thermodynamic Calculation of the Al-N-O System," *Z. Metallkd.*, **83**[10], 714-19 (1992).

[12] C. Qui and R. Metselaar, "Phase Relations in the Aluminum Carbide-Aluminum Nitride-Aluminum Oxide System," *J. Am. Ceram. Soc.*, **80**[8], 2013-20 (1997).

[13] W. Nakao, H. Fukuyama, and K. Nagata, "Thermodynamic Stability of $\gamma$-Aluminum Oxynitride," *J. Electrochem. Soc.*, **150**[2], J1-J7 (2003).

# LARGE AREA EFG™ SAPPHIRE FOR TRANSPARENT ARMOR

Christopher D. Jones, Jeffrey B. Rioux, John W. Locher, Herbert E. Bates, and Steven A. Zanella
Saint-Gobain Crystals
33 Powers Street
Milford, NH, 03055

Vincent Pluen, and Mattias Mandelartz
Saint-Gobain Sully
16 route d'Isdes
Sully sur Loire, France

ABSTRACT

Edge Defined Film-fed Growth (EFG™) sapphire crystals are being grown commercially as large, thick sheets that measure greater than 305 mm wide, 775 mm long, and 11 mm thick. The high strength, high hardness, and good transparency in the visible and infrared spectra make sapphire an ideal choice for transparent armor applications. This paper will discuss the properties of large sapphire sheets and the results from ballistic testing of selected sapphire transparent armor panels.

INTRODUCTION

Large Saphikon® EFG™[1] sapphire sheet measuring 225 x 323 x 10 mm have been in production since 2003 and have been reported on previously.[2-6] Over 1000 crystals of this size have been produced to date. However the demand for even larger area window material in the VIS-MWIR (500 to 5000 nm) spectrum is growing and has served as the rationale for scaling up to the larger sizes presented here. In the past three years (2003-2005) redesigned hot zones and equipment have produced crystals 305 x 510 mm and 225 x 660 mm in dimensions. Saint-Gobain Crystals now offers their CLASS[225] and CLASS[300] sapphire in production quantities. The largest application of these windows to date has been in the aerospace industry for VIS-MWIR optical windows, where the intrinsic strength, hardness (resistance to erosion), chemical resistance, and high optical transmission make sapphire a highly desirable material. These same properties are also highly desirable for transparent armor. As threat levels increase, so does the need for stronger and superior transparent armor systems. Current military transparent armor is made from laminated glass. The thickness requirements to stop certain rounds lead to a cumbersome window style that has reduced transmission. Additionally, it can be difficult to use side mirrors and the added weight can lead to hinges failing and reduced agility. Ceramic composites are one way to improve upon traditional transparent armors such as glass and laminated glass. The use of a ceramic strike-face in a transparent armor system allows one to take advantage of the high strength and hardness of a ceramic, usually with a weight and thickness advantage over the equivalent glass-based system. This paper will discuss the characteristics of EFG™ sapphire that make it a leading material for transparent armor systems. Herein we report the first ballistic tests on sapphire transparent armor systems.

SAPPHIRE

Growth of Sapphire Strike-Faces

Sapphire, or single crystal aluminum oxide in the corundum crystal structure, can be grown by a variety of methods. Common industrial techniques include the Czochralski, Kyropolous, Heat-Exchanger Method (HEM), Bridgman, and Edge-defined Film-fed Growth (EFG™) techniques. While the authors will not debate the various growth techniques in this paper (an excellent review can be found in reference [7]), we do acknowledge that the Edge-defined Film-fed Growth technique allows for surfaces areas approximately two-fold larger than the competitive techniques.

While EFG™ has been used industrially for several decades, demand has driven larger sapphire windows in the past 5 years. Figure 1 shows an illustration of the Saphikon® EFG™ technique. Aluminum oxide is melted in a crucible using induction heating. A die that contains a thin capillary leading to a shaped die tip is placed into the melt. Due to the melt wetting the die, the melt rises to the die tip through capillary forces. A seed with a specific crystallographic orientation is placed into contact with the melt and pulled away, allowing the melt to crystallize into sapphire. The die tip may have any cross section (sheet, rod, tube, curved sheets, etc.).

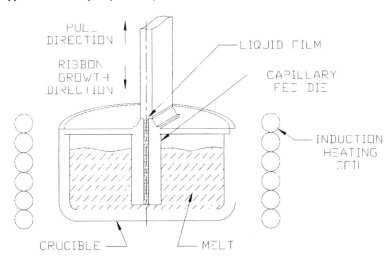

Figure 1. Illustration of the Edge-defined Film-fed Growth (EFG™) technique.

As-grown 225 x 660 mm and 305 x 510 mm crystals are shown in figure 2. These crystals are then processed to remove their surfaces, annealed and polished to form the final windows shown in figure 3. Changing the thickness or cross-section of the die can produce various thicknesses and cross-sections of the sapphire.

Figure 2. As-grown CLASS²²⁵™ (225 x 660 mm) and CLASS³⁰⁰™ (305 x 510 mm) EFG™ sapphire crystals

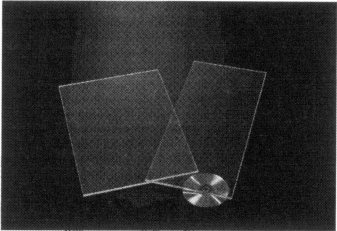

Figure 3. Polished CLASS²²⁵™ and CLASS³⁰⁰™ EFG™ sapphire crystals.

Ceramic Materials Physical Properties

The four major strike-face materials used in armor are glass, sapphire ($Al_2O_3$), aluminum oxynitride [$AlN_x(Al_2O_3)_{1-x}$], and spinel ($MgAl_2O_4$). Table I contains pertinent material properties for each of these materials. Sapphire has the best fracture toughness, flexure strength, thermal conductivity, hardness and modulus of elasticity (Young's modulus) of all the strike face materials. The density of sapphire is almost double glass, but only slightly higher than the other ceramic-based strike-face materials. Based on the mechanical properties, sapphire is expected to have the best ballistic performance of all the strike-face materials for transparent armor.

Table I. Material Properties of Transparent Armor Materials[8,9,10]

| | Material | Sapphire | Aluminum oxynitride | Spinel | Glass |
|---|---|---|---|---|---|
| Chemical Formula | | $Al_2O_3$ | $(AlN)_x \cdot (Al_2O_3)_{1-x}$, $0.30 \leq x \leq 0.37$ | $MgAl_2O_4$ | $SiO_2$ |
| Classification | | single crystal | polycrystalline | polycrystalline | amorphous |
| Process Method | | melt growth | HP/HIP | HP / HIP | float |
| Transmission | microns | 0.2-6.0 | 0.2 to 5.5 | 0.2 to 6.5 | 0.2-4 |
| Fracture Toughness | MPa-m$^{1/2}$ | 2 | 1.4 | 1.9 | 0.9 |
| Flexure Strength | MPa | 750 | 300 | 190 | 69 |
| Knoop Hardness | GPa | 22 | 18 | 14.9 | 4 |
| Thermal Conductivity | (W/m·K) | 34 | 12.6 | 14.6 | 1.3 |
| Density | g/cc | 3.97 | 3.69 | 3.59 | 2.47 |
| Young's Modulus | GPa | 345 | 323 | 275 | 10 |
| Poisson's Ratio | | 0.29 | 0.239 | 0.26 | 0.2 |

SAPPHIRE TRANSPARENT ARMOR SYSTEM

We report herein the first results of our transparent armor composite using Saphikon® EFG™ sapphire and glass. The sapphire strike-face in all of the composite armor reported here is 6.35 mm thick. The sapphire is bonded to glass using conventional interlayers, such as polyvinyl butyral and polyurethane. While the exact composite-make up is proprietary, each composite has a polyurethane backing designed to contain spalling.

Ballistic Tests

Each sapphire transparent armor system was tested using a 150x150mm sample and ballistically tested using a single shot. The sapphire sheets were grown and processed by Saint-Gobain Crystals and Saint-Gobain Sully did the assembly of the sapphire sheets with a further armored glass make-up including a polycarbonate sheet on the inside in order to prevent spalling. The composite physical properties and ballistic results are shown in Table II. Due to the existence of multiple ballistic standards, we will not report here on meeting specific levels of standards, but rather report on the projectiles and velocities within each test. A partial penetration means that the armor strike-face was penetrated, however the projectile and all of the spalling was contained by the polycarbonate layer (the final polycarbonate layer was not penetrated). A complete penetration means that the spalling went through the final polycarbonate layer (the final polycarbonate layer was penetrated).

The ballistic tests were performed under controlled conditions at certified ballistic ranges. For sample ID# 1-8, the projectile was fired at 10m and optical velocity detector was used to determine the projectile velocity 2.5m in front of the target. For sample ID# 9-11, the projectiles were fired at 50m and an optical velocity detector was used to determine the projectile velocity at 6.5m in front of the target.

Table II: Saint-Gobain Sapphire-Glass Composite Armor Ballistic Results

| Sample ID # | Length (mm) | Width (mm) | Thickness (mm) | Mass (kg) | Areal Density (kg/m²) | Projectile | Projectile Velocity (m/s) | Penetration* |
|---|---|---|---|---|---|---|---|---|
| 1 | 150 | 150 | 21.1 | 1.22 | 52.12 | 7.62x51 M-80 Ball | 835 | Partial |
| 2 | 150 | 150 | 21.1 | 1.22 | 52.12 | 7.62x39 API-BZ | 696 | Partial |
| 3 | 150 | 150 | 21.1 | 1.22 | 52.12 | 7.62x39 API-BZ | 746 | Partial |
| 4 | 150 | 150 | 21.1 | 1.22 | 52.12 | 7.62x39 API-BZ | 776 | Partial |
| 5 | 150 | 150 | 21.1 | 1.22 | 52.12 | 7.62x51 AP(M61) | 768 | Partial |
| 6 | 150 | 150 | 21.1 | 1.22 | 52.12 | 7.62x51 AP(M61) | 846 | Complete |
| 7 | 150 | 150 | 21.1 | 1.22 | 52.12 | 7.62x51 AP(M61) | 864 | Complete |
| 8 | 150 | 150 | 21.1 | 1.22 | 52.12 | 7.62x51 AP(M61) | 867 | Complete |
| 9 | 150 | 150 | 29.4 | 1.63 | 72.78 | 7.62x54R B32 | 857 | Partial |
| 10 | 150 | 150 | 29.4 | 1.63 | 72.78 | 7.62x54R B32 | 858 | Partial |
| 11 | 150 | 150 | 29.4 | 1.63 | 72.78 | 7.62x54R B32 | 855 | Partial |

*In ballistic testing, partial penetration is considered a passing result.

The first samples tested (Sample ID #1 through 8, see Table II) were 21.1mm thick and had an areal density of 52.12 kg/m². This composite make-up defeated the 7.62x51mm M80 ball at a velocity of 835 m/s (Sample ID#1). The deformation of this transparent armor system may be seen in figures 4 and figure 5. Using this same configuration (Sample ID# 2-4), it was then tested using the 7.62x39mm API-BZ. The system also defeated this round at three different velocities, up to 776 m/s. The deformation of the 776m/s round is shown in Figures 4 and 5. It can be noted that the 7.62x51mm M80 round had a larger deformation, most likely due to the higher momentum M80 round compared to the API-BZ.

Testing of the 7.62x51mm AP M61 (hard steel core) round was performed on the 21.1 mm thickness (Sample ID# 5-8, see Table 2). While the round was defeated (partial penetration) at 768 m/s, this is slower than normal testing velocity. When tested at normal test velocities (above 820 m/s), all three tests resulted in complete penetration. It is believed that a slight change in the glass composition will allow a slightly thicker system to also defeat this round.

The next test system subjected to ballistic testing was slightly thicker (Sample ID#9-11, see Table II) at 29.4mm and an areal density of 72.78 kg/m². This system make-up defeated (partial penetration) the 7.62x54R B32 API round at a velocity of 858 m/s. For comparison, a glass-only solution to defeat this round would be over 50 mm thick and have an areal density greater than 129 kg/m².

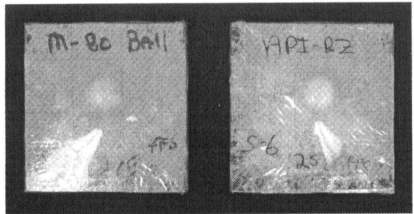

Figure 4. Sapphire transparent armor system after being hit by the 7.62x51 M-80 Ball (left) at 835 m/s and by the 7.62x39 API-BZ at 776 m/s (right) showing shot placement.

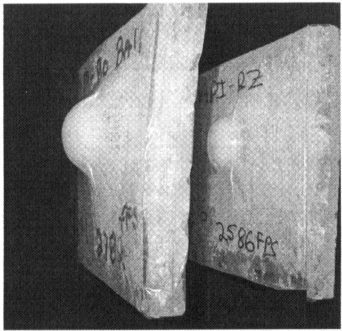

Figure 5. Sapphire transparent armor system after being hit by the 7.62x51 M-80 Ball (left) at 835 m/s and by the 7.62x39 API-BZ at 776 m/s (right) showing displacement.

Optical Tests

A significant advantage of using sapphire as a strike-face is the high optical transmission in the visible and infrared spectrum. A sapphire transparent armor system at 29.4 mm thickness has a transmission greater than 85%, with haze levels around 1%, see Table III. Increasing the thickness of the system by adding additional glass layers to 41.1mm thickness only reduces the transmission to around 84%, with no change in haze. For comparison, a typical glass window that has similar ballistic protection as the 29.4 mm sapphire transparent armor system has a luminous transmission of 73% and haze around 0.6%.

Table III: Optical Measurements on Saint-Gobain EFG™ sapphire transparent armor system

| Thickness (mm) | Areal Density (kg/m$^2$) | Luminous Transmission (%) | Haze (%) |
|---|---|---|---|
| 29.4 | 72.78 | 85.9 | 0.99 |
| 29.4 | 72.78 | 85.3 | 1.25 |
| 41.1 | 101.11 | 84.4 | 1.00 |
| 41.1 | 101.11 | 84.3 | 1.12 |

FUTURE WORK

Given the exciting results presented in this paper, Saint-Gobain is continuing to develop and invest in the ceramic transparent armor market. We are currently investigating the impact of ballistic testing on thicker ceramic sheets and system design. We have prepared larger sapphire transparent armor system samples capable of testing multiple hits, see Figure 7. The Saint-Gobain EFG™ technology has a benefit that it is scalable to larger sizes with only design and procurement of larger capital equipment. Sapphire transparent armor systems will benefit from the scale up for larger optical material requirements. To address the market requirements for larger optical materials, CLASS$^{400™}$ will be available in the fall of 2006.

CONCLUSIONS

Sapphire has excellent mechanical and physical properties as a strike face for transparent armor. Saint-Gobain has shown that sapphire transparent armor systems are able to have good ballistic performance with thinner windows, lower areal density, and greatly improved transmission over traditional transparent armors. It has been shown that a sapphire transparent armor systems with an overall thickness of 21.1mm and an areal density of 52.1 kg/m$^2$ could defeat the 7.62mmx51 M80 Ball at velocity of 835 m/s, the 7.62mmx29 API-BZ at velocity 776m/s, and the 7.62mmx51 AP(M61) at velocity 768 m/s. A 29.4mm thick sapphire transparent armor system with areal density of 72.8 kg/m2 was also shown to defeat the 7.62x54R B32 projectile at a velocity of 858 m/s. The sapphire transparent armor system is at least 40% lighter and 40% thinner than the equivalent glass system required to defeat the 7.62x54R B32 projectile. Given these excellent results, Saint-Gobain is continuing to develop sapphire as a transparent armor solution.

Figure 7: A 305x305mm sapphire transparent armor system for multi-hit testing.

ACKNOWLEDGEMENTS
The authors wish to thank Brett Elkind and Rich D'eon at Protective Materials Company, a division of The Protective Group for their assistance with some of the ballistic tests.

REFERENCES
1. H. E. LaBelle, *EFG the invention and application to sapphire growth*, Journal of Crystal Growth, Vol. 50, 8 (1980).

2. J.W. Locher, H.E. Bates, S.A. Zanella, E.C. Lundstedt, and C.T. Warner, *The production of 225 x 325 mm sapphire windows for IR (1 to 5 µm) application*, Proc. SPIE Vol. 5078, Window and Dome Technologies VIII. Randal W. Tustison, Editor, 2003, pp.40-46

3. J. W. Locher, J. B. Rioux, and H. E. Bates, *Refractive Index Homogeneity of EFG A-plane Sapphire for Aerospace Windows Applications*, 10th DoD Electromagnetic Windows Symposium, Norfolk VA, 18-20 May 2004.

4. H. E. Bates, C. D. Jones, and J.W. Locher, *Optical and crystalline characteristics of large EFG Sapphire Sheet*. Proc. SPIE Vol. 5786, Window and Dome Technologies and Materials IX. Randal W. Tustison, Editor, May 2005, pp. 165-174

5. J. W. Locher, H.E. Bates, C. D. Jones, and S. A. Zanella, *Producing large EFG sapphire sheet for VIS-IR (500-5000 nm) window applications*, Proc. SPIE Vol. 5786, Window and Dome Technologies and Materials IX, Randal W. Tustison, Editor, May 2005, pp. 147-153

6. C. D. Jones, J. W. Locher, H.E. Bates, and S. A. Zanella, *Producing large EFG sapphire sheet for VIS-IR (500-5000 nm) window applications*, Proc. SPIE Volume 5990, 599007 (2005).

7. V. A. Tatartchenko, *Sapphire Crystal Growth and Applications*, in Bulk Crystal Growth of Electronic, Optical, and Optoelectronic Materials, edited by P. Crapper, pp 299-338, John Wiley & Sons (2005)

8. D.C. Harris, *Materials for infrared windows and domes, properties and performance*, SPIE Optical Engineering Press, Bellingham WA, 36 (1999).

9. Surmet Corporation Company Literature #M301030

10. Saint-Gobain Crystals Product Literature

# Other Opaque Ceramics

# RELATIONSHIP OF MICROSTRUCTURE AND HARDNESS FOR $Al_2O_3$ ARMOR MATERIALS

Memduh Volkan Demirbas
Rutgers University
Ceramic and Materials Engineering
607 Taylor Road
Piscataway, NJ 08854

Richard A. Haber
Rutgers University
Ceramic and Materials Engineering
607 Taylor Road
Piscataway, NJ 08854

ABSTRACT

In armor ceramics, it is hard to separate "good" from "bad", with samples that have close density values. It is still unknown whether a slight change in residual porosity above 99.5%, namely the shape and dispersity of pores are detrimental to the dynamic performance. The method for solving this problem is based on spatial data analysis and subsequent mechanical tests. Three hot-pressed armor grade $Al_2O_3$ ceramic tiles were used for this purpose. Microstructural assessment was carried out using nearest neighbor distance distribution functions and tessellation analysis. Hardness tests were performed to assess the properties of the samples. Hardness maps were obtained by indentation of samples 100 times, forming a square array of $10 \times 10$ indents and presented as contour maps. The results from microstructural assessment and hardness tests were compared and some degree of correlation was observed between the microstructural analysis and hardness tests.

INTRODUCTION

In armor materials, the origin of microstructural anomalies and their effect on the properties of bulk ceramics is a key issue. Thus, microstructural assessment can provide useful tools to understand the development of microstructure and resulting materials properties. However, rarely are results from microstructural analysis compared with mechanical properties. Although ballistics tests would be the ultimate solution to this problem, a more convenient and easy way would be preferred. It was reported in the literature that several static properties, such as hardness, sonic velocities, Young's modulus, Poisson's ratio and density have an effect on ballistic performance. Hardness has been found to be the only property that is helpful in predicting the ballistic performance of a material[1]. It is also a relatively convenient testing method so hardness tests and their statistical interpretation were used in this study.

For microstructural assessment part, the emphasis in this paper is given to pores and especially their spatial and planar distribution in the material. This concept is important since the variations in density in a sample might be detrimental to the mechanical properties. Microstructural analysis is a reliable method to assess the nature of variations in density from sample to sample.

For the microstructural assessment part, two similar techniques, namely nearest neighbor distance distribution and tessellation analysis, were used. Nearest neighbor distance distributions are obtained by measuring the distance between a pore and its closest neighbor, then this is done for all the pores present in the micrograph and a distribution is obtained at the end[2, 3].

Tessellation analysis is the geometric construction of two dimensional cells or polygons around each feature on the plane-of-polish. This allows for a quantitative assessment of the microstructure, as the cell area is dependent on the whole surrounding environment of the pore concerned[4, 5]. It is a similar approach to nearest neighbor distance distributions since both are concerned with the surroundings of each pore in the field of view. Details of microstructural analysis were explained by the authors previously[6].

## EXPERIMENTAL PROCEDURE

Three commercial armor grade Al$_2$O$_3$ samples from a single production lot were used throughout this study. Grinding was performed using 125, 40, 20, 10 μm diamond grits and they were polished with 6 and 1 μm diamond suspensions. A LECO tester was employed at a load of 2 kg for 10 seconds to obtain Knoop hardness values. Knoop indenter was used in this study instead of Vickers in order to avoid cracking.

The regions were selected randomly for microhardness measurements. One hundred indents were taken from each sample forming a square from 10×10 indents. 0.5 mm distance was left between each indent. It was intended to place the indents so that they did not interfere with each other.

Image analysis was performed on the micrographs from polished sections to obtain quantitative information on the samples. First step of the image analysis was thresholding, which was performed to obtain a binary image. This was followed by closing operation to smooth features and to remove isolated pixel noise from the image. Filling holes was performed after that and the final step was cutting off edge-touching features in order to include only the features in the field of view[7, 8]. Finally, a report on the features is obtained, which provides information such as equivalent diameter, x- and y- coordinates of each feature in the image for further analysis.

For the estimation of nearest neighbor distance distributions, the x- and y- of the pore centroid coordinates must be measured on the ceramographic plane. Let m be the total number of pores in the field of view and the centroid coordinates be $(x_1,y_1)$, $(x_2,y_2)$, . . ., $(x_i,y_i)$, . . ., $(x_m,y_m)$. Then, the distances are calculated between each pore and all other (m-1) pores. The nearest- (or higher order) neighbor distance for each pore is calculated from this information and a frequency histogram of the nearest- (or higher order) neighbor distances is obtained[8, 9].

The second technique that was used in this study was tessellation analysis. The first step in the tessellation analysis is thresholding, to obtain a binary image with a pixel value either 0 or 1. Then the image is inverted to reverse the pixel values. Skeletonization is performed afterwards to extract a region-based feature representing the general form of an object. The image is inverted back again to quantify the cells that are formed after the previous steps. The sequence of events explained here is shown in Figure 1.

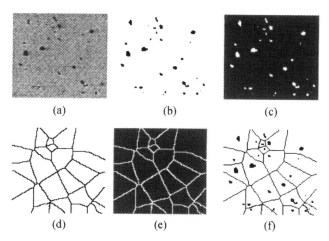

Figure 1. (a) Original image (b) Thresholding (c) Inversion (d) Skeletonization (e)Inversion (f) Combined image of cells and actual pores

STATISTICAL ANALYSIS

In this experimental investigation, Knoop microhardness tests were used to assess the heterogeneity of armor materials. Statistical elaboration of obtained results was performed in order to separate variation due to scatter of data. Statistical analysis of the data is also important so as to quantify the spread of data since same type of information will be obtained from spatial data analysis and this will create a comparison basis between the two sets of data.

Property distributions may be highly skewed or broadly distributed, and this was case in our data set, therefore, Weibull distribution was used to statistically analyze the hardness data sets. The Weibull distribution, in the two-parameter form, is given as:

$$F(x) = 1 - \exp\left[-\left(\frac{x}{x_0}\right)\right]^m \qquad (1)$$

where, F(x) is the cumulative density function of probability, x is the microhardness data, $x_0$ is the scale parameter below which 63.2% of the data lie, and m is the Weibull modulus. It is basically the parameter, which reflects the data scatter within the distribution[10, 11].

Weibull plot is commonly used to obtain Weibull parameters. Equation (1) is rearranged and gets into the following form:

$$\ln\left\{\ln\left[\frac{1}{1-F(x)}\right]\right\} = m[\ln(x) - \ln)x_0] \qquad (2)$$

One issue here is to determine the way approach to calculate F(x) values. Several different estimators have been used, the most common one being:

$$F(x) = \frac{i}{n+1} \tag{3}$$

where, n is the total number of data points, and $i$ is the ith order in ascending data set. For highly skewed distributions, which were the case in this study, the following estimator has been used [10-13].

$$F(x) = \left( \frac{i - 0.3}{n + 0.4} \right) \tag{4}$$

RESULTS

The micrographs of three individual samples are given in Figure 2. The microstructures look similar, although there are some minor differences visually. Sample A seems to have some individual larger pores compared to the other two samples and more elongated shaped pores are present. There are dense areas and small clusters of pores in the micrographs of Sample A and B whereas the pores seem to be more evenly distributed throughout the sample for Sample C. These observations will be quantified in this section.

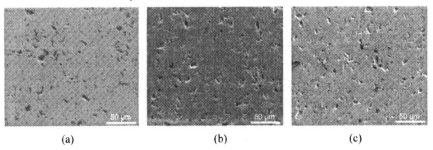

(a)     (b)     (c)

Figure 2. Micrographs of (a) Sample A. (b) Sample B, (C) Sample C

The densities of the samples were calculated using image analysis and are 3.86, 3.87, 3.90 g/cm$^3$ for A. B, and C. respectively. Density values are given in Figure 3.

Average pore size and pore size distribution of these samples were then calculated from image analysis. Equivalent diameter of the features was used to calculate average pore diameter.

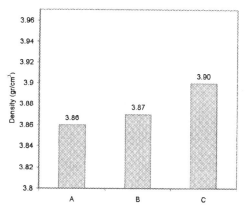

Figure 3. Density bar chart for the three samples

In Figure 4(a), the pore size distribution graphs are given. Sample C has largest peak in 0.5-1.0 μm range, with almost 50% of the pores in this range. For the same range, Samples B and C have approximately 45% and 40% of the total number of pores, respectively. For the next size range, 1.0-1.5 μm, the trend is reversed. It can be concluded that among the three samples, Sample C has the finest pores. This is also shown in the average pore size chart in Figure 4(b). Samples A, B, and C have an average pore size of 1.69 ± 1.50 μm, 1.45 ± 1.24 μm, and 1.38 ± 1.14 μm, respectively.

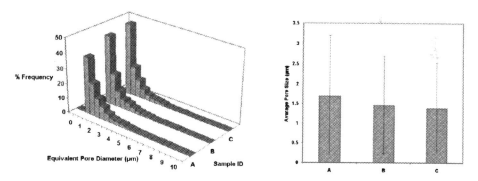

Figure 4. (a) Average pore size distribution (b) Average pore size graph by image analysis

Nearest neighbor distance distributions were calculated to learn about the spatial distribution of pores in the material, rather than the information only about size. After the distribution graphs were obtained, the observed mean and variance values were compared to that of a random Poisson distribution[2]. The nearest neighbor distance distribution graph is given in Figure 5. The nearest neighbor distance distributions are relatively similar for Samples B and C, with the major peaks in 0-4 μm range and their percentages approximately 20%. However, there is a shift in the distribution for Sample A, with the major peaks in 4-8 μm range and the tail extends unlike the curves for the other two samples. This is a clear indication of less homogeneous microstructure that this sample possesses.

The average nearest neighbor distance *(nnd)* values for each sample from A to C is given as $7.58 \pm 3.83$ μm, $5.17 \pm 2.98$ μm, and $5.62 \pm 3.32$ μm, respectively. These values are compared with the ones calculated by Image Processing Toolkit 3.0 for a random distribution. The nature of a distribution may be described according to two parameters Q and V, which are obtained by the following formulas[2]:

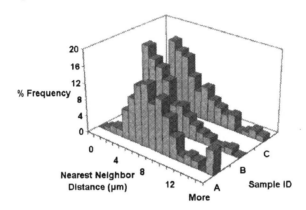

Figure 5. Nearest neighbor distance distributions of Al₂O₃ samples

$$Q = \frac{\mu_o}{\mu_e} = \frac{\text{observed mean nearest - neighbor distances}}{\text{expected mean nearest - neighbor distances}} \qquad (5)$$

$$V = \frac{\text{var}_o}{\text{var}_e} = \frac{\text{observed variance of nearest - neighbor distances}}{\text{expected variance of nearest - neighbor distances}} \qquad (6)$$

The closer these two parameters are to 1, the closer they are to random distribution case. The values for the alumina samples are given in Table I.

Table I. Q and V table

|   | Mean nnd | Variance nnd | Mean nnd (random) | Variance nnd (random) | Q | V |
|---|---|---|---|---|---|---|
| A | 7.58 | 14.68 | 6.67 | 1.73 | 1.14 | 8.05 |
| B | 5.17 | 8.86 | 5.08 | 2.27 | 1.02 | 3.91 |
| C | 5.62 | 11.02 | 5.93 | 2.65 | 0.95 | 4.16 |

Q-V chart plotted from the values above is given in Figure 6. If the random point for a distribution is accepted as the point (1,1) in the graph, which is the intersection of the axes, then the closer point to (1,1) must be considered as the most random case. The distances to (1,1) were calculated using $(Q^2+V^2)^{1/2}$ formula and they came out as 8.13, 4.04 and 4.27 units for Sample A, B and C, respectively.

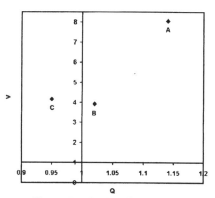

Figure 6. Q-V chart for the Al$_2$O$_3$ samples

Tessellation analysis was the other technique applied to assess microstructures. Cell area distributions from tessellation analysis for alumina samples are given in Figure 7. Sample B has a relatively narrower distribution than the other two, with the major peaks reaching 15% of all cells and the tail representing larger cells has considerably smaller values. Sample A follows with the major peaks approximately 12%, and the peaks in the tail have slightly larger values. Sample C has clearly the broadest distribution among the three, with a significant amount of cells larger than 200 $\mu m^2$ cell area.

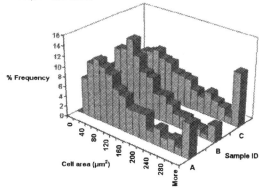

Figure 7. Cell area distributions

Two parameters, P$_1$ and P$_2$, were defined by Murphy *et al.*[?] to describe the nature of a distribution using tessellation analysis instead of nearest neighbor distance distributions. P$_1$ and P$_2$ are defined as:

$$P_1 = \frac{V}{V_{random}} \qquad (7)$$

$$P_2 = \frac{S}{S_{random}} \qquad (8)$$

where, $V$ is the variance of measured cell areas, $V_{random}$ is the variance of a random distribution, $S$ is the skewness of measured cell areas and $S_{random}$ is the skewness of a random distribution. P$_1$ and P$_2$ chart is given in Figure 8.

Sample A has the largest P$_1$ and P$_2$ values, 1.18 and 1.53, respectively, as opposed to 1, which is the expected value for random distribution of particles. P$_1$ for Sample B is almost 1 with 1.01 value, however P$_2$ value is much smaller than 1. According to those values, Sample B and C have pore distributions closer to random case than it is for Sample A. In this case, the results are similar to those of nearest neighbor distance distributions.

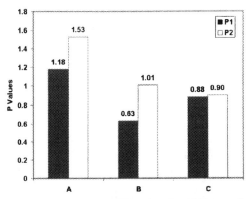

Figure 8. P$_1$ and P$_2$ values for Al$_2$O$_3$ samples

Hardness maps were constructed to see the variation in hardness values terms of location. One hundred indentations were taken from each sample and a square was formed from 10×10 indentations. There are 500 μm differences between each indent. Shorter distances were also tried but the indents became too close and interfered with each other so 500 μm turned out to be a reasonable value. Possible low hardness regions might be detrimental to the overall performance of the material so these hardness maps are important in monitoring the variations. Hardness maps of the samples are given in

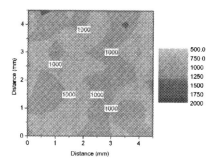

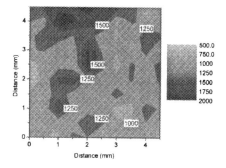

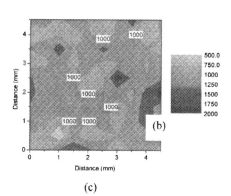

(c)

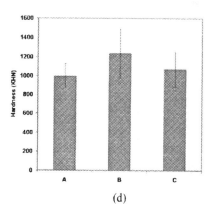

(d)

Figure 9. (a) Hardness map of Sample A (b) Hardness maps of Sample B (c) Hardness maps of Sample C (d) Average hardness values

Figure 9(a)-(c) and the average hardness values are given in Figure 9(d). All values shown on the maps are Knoop hardness values (KHN). Figure 9(d) shows that Sample B has the highest average hardness value, Sample C comes second and Sample A has the lowest average value. However, Sample A shows a more uniform and homogenous distribution, which is apparent from the maps. Homogeneity is less pronounced for the Samples B and C, with a larger spread between low and high hardness values.

Weibull analysis was performed on 100 hardness indents. Weibull plot for all the samples is given in Figure 10. As explained previously, Weibull data shows the extent of spread of a distribution and observations made from the hardness maps are quantified by using Weibull analysis. Sample A has the highest Weibull modulus with 9.14, followed by Sample C with 6.88 and Sample B with 5.91.

The last part of this study was to determine whether microstructural data and hardness data directly correlated with each other. Hardness vs. density and hardness vs. average pore size graphs are given in Figure 11(a) and 11(b), respectively.

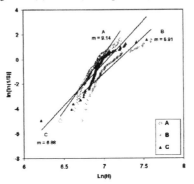

Figure 10. Weibull plot for the three Al₂O₃ samples

Both graphs show a similar type of curve due to close correlation between density and average pore size. No strong link is apparent in these graphs. However a reasonable trend is still observable if a trend line is drawn, which is, samples with higher densities and smaller average pores sizes possess higher hardness values.

Weibull modulus values were plotted against the numbers obtained from nearest neighbor distance distributions, given in Figure 12. The smallest number from *nnd* distributions is from Sample B. For the same sample, highest Weibull modulus value was expected for Sample B in order for the microstructural data to match with hardness data, however this was not the case.

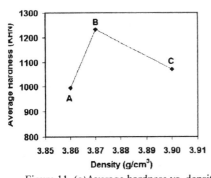
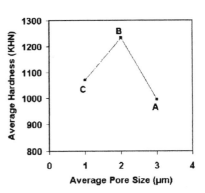

Figure 11. (a)Average hardness vs. density
(b) Average hardness vs. average pore size

The positive slope in the graph should have been negative in order to obtain a close correlation. This was not observed for these particular samples. The reason behind this is mainly the larger spread in hardness data for Sample B and the narrow distribution of hardness values for Sample A, although there is considerable difference between average values.

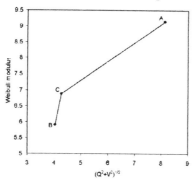

Figure 12. Weibull modulus vs. $(Q^2+V^2)^{1/2}$

Another correlation between spatial microstructural data and hardness is given Figure 13, where P parameters were plotted against average hardness. The highest point of hardness value was expected to reach when P values were close to 1, if random distribution of pores was favored. This was valid for $P_2$, which is based on the skewness of cells in tessellation analysis. However, same correlation was not observed for $P_1$, with Sample B standing out with a high hardness value but a low $P_1$.

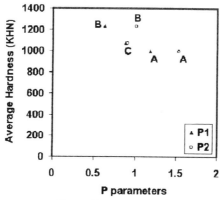

Figure 13. Average hardness vs. $P_1$&$P_2$

CONCLUSION

Three different $Al_2O_3$ samples were investigated in terms of microstructural uniformity and hardness. Microstructural assessment was performed using nearest neighbor distance distributions and tessellation analysis. Hardness tests were utilized to obtain hardness maps, which show hardness data in terms of location. Hardness was employed in this study since it has been the only materials property that helps to predict ballistic performance. Weibull analysis was also used to quantify the spread in the hardness data. The information from both analyses was used to draw a conclusion for microstructural uniformity and its effect on the quasi-static properties. For this particular case, hardness increases with higher density and lower average pore size. There is also some evidence from tessellation analysis that homogenous distribution of pores has a positive effect on hardness data. However, different set of samples will be utilized to strengthen this hypothesis.

REFERENCES

[1]D. Viechnicki, W. Blumenthal, M. Slavin, C. Tracy and H. Skeele, "Armor Ceramics – 1987", The Third Tacom Armor Coordinating Conference Proceedings

[2]J. P. Anson, J. E. Gruzlezski, "The Quantitative Discrimination between Shrinkage and Gas Microporosity in Cast Aluminum Alloys Using Spatial Data Analysis", Materials Characterization 43, 319-335 (1999)

[3]Joachim Ohser, Frank Mucklich, "Statistical Analysis of Microstructures in Materials Science", 2000 John Wiley & Sons, Ltd.

[4]George F. Vander Voort, "Computer-Aided Microstructural Analysis of Specialty Steels", Materials Characterization 27, 241-260 (1991)

[5]A. M. Murphy, S. J. Howard, T. W. Clyne, "Characterization of severity of particle clustering and its effect on fracture of particulate MMC's", Journal of Materials Science 31 (20), 5399-5407 (1996)

[6]M. V. Demirbas and R. A. Haber, "Defining Microstructural Tolerance Limits of Defects for SiC Armor", 107th Annual Meeting and Exposition of the American Ceramic Society, April 11-13, 2005, Baltimore, MD.

[7]P. Louis, A. M. Gokhale, "Application of image analysis for characterization of spatial arrangements of features in microstructures", *Metallurgical and Materials Transactions A*, Vol. 26A, 1449-1456, June 1995

[8]John C. Russ, "Computer-Assisted Microscopy", 1990 Plenum Press, New York

[9]M. Berman, L. M. Bischof, E. J. Breen, G. M. Peden, "Image Analysis', Materials Forum 18, 1-19 (1994)

[10]Jianfeng Li, Chuanxian Ding, "Determining microhardness and elastic modulus of plasma sprayed $Cr_3C_2$-NiCr coatings using Knoop indentation testing", Surface and Coatings Technology 135 (2001) 229-237

[11]C. K. Lin, C. C. Berndt, "Statistical analysis of microhardness variations in thermal spray coatings", Journal of Materials Science 30 (1995) 111-117

[12]B. Bergman, "On the estimation of Weibull modulus", Journal of Materials Science Letters 3 (1984) 689

[13]M. T. Lin, D. Y. Jiang, L. Li, Z. L. Lu, T. R. Lai, J. L. Shi, "The effect of creep deformation of a β-Sialon on Vickers hardness, fracture toughness and Weibull modulus", Materials Science and Engineering A 351 (2003) 9-14

# ROOT CAUSES OF THE PERFORMANCE OF BORON CARBIDE UNDER STRESS

Giovanni Fanchini 1, Dale E. Niesz 1, Richard A. Haber 1, James W. McCauley 2, and Manish Chhowalla 1

1 Materials Science and Engineering, Rutgers University
Piscataway, NJ 08854, U.S.A.
2 ARL, Aberdeen Proving Ground
Aberdeen, MD 21005, U.S.A.

## ABSTRACT

The absence of a plastic phase in boron carbide and its failure at shock impact velocities just above the Hugoniot elastic limit ($H_{EL}$) has been the subject of several experimental investigations. Furthermore, the common presence of contaminants, such as disordered graphitic inclusions, oxygen, etc., needs to be addressed. Further, a theoretical picture accounting all these phenomena is still lacking. In the present work, using self-consistent field density functional simulations we are able to account for many experimental observations by noticing that several boron carbide polytypes [e.g. $(B_{11}C)CBC$, $(B_{12})CCC$, ...] coexist without significant lattice distortions. Our analysis also indicates that above a threshold pressure all such polytypes are less stable than a phase involving segregated boron ($B_{12}$) and amorphous carbon (a-C) but the energy barrier for the transformation into a segregated phase of boron and carbon, is by far lower for the $B_{12}(CCC)$ polytype. For such a configuration, segregation of carbon occurs in layers orthogonal to the (113) lattice directions, in excellent agreement with recent transmission electron microscopy (TEM) analysis. We will also, in the actual preparation conditions of the material, show that the Gibbs free energy per site in the $B_{12}$ + a-C segregate phase, in $B_{12}O_2$ and $B_4C_{1-x}Si_x$ is not significantly lower than in most of the $B_4C$ polytypes. Silicon inclusions, however, should strongly reduce the formation of the $(B_{12})CCC$ phase.

## INTRODUCTION

Boron carbide has a unique combination of properties: hardness (H = 35 GPa) low density ($\rho$ = 2.51 g cm$^{-3}$) and the highest dynamic elasticity in ceramics (Hugoniot elastic limit $H_{EL}$ = 17-20 GPa) with the possibility of massive production by hot pressing. This makes it the material of choice for several key applications, such as abrasive powders, neutron control rods and lightweight armors [1]. The performances of boron carbide as an armor material rely on the persistence of elastic behavior at threat velocities up to 850 m s$^{-1}$ surpassing by a factor two all of its competitors, such as SiC and Al$_2$O$_3$, which are, furthermore, 50% denser [2]. However SiC and AlO$_2$ when stressed above the elastic limit exhibit a residual plastic behavior, allowing protection against high velocity threats at the cost of little permanent damage. In contrast, the use of boron carbide in such conditions is not possible now because the material catastrophically fails at dynamic pressures just above the $H_{EL}$ [1]. The physics of such a phenomenon [1], a unique behavior in dynamically elastic materials, is unclear, whilst similar effects can be found, although at much lower stresses (<1 GPa), only in dynamically inelastic materials, such as quartz and glasses.

Recently, some efforts have proven which the causes of the failure are not. Voegler *et al*

[3] show that a phase transition can be found only at 40 GPa pressure or above, a too high value to explain the observed failure. A phase transition explaining the failure at the uniaxial stresses comparable to $H_{EL}$ must occur at hydrostatic pressures $P \approx 7\text{-}8$ GPa, corresponding to the pressure P(HEL) occurring on the hydrostatic stress-strain curve at the Hugoniot conditions [2]. On the other hand, Chen $et\ al$ [4] proved that the reasons of the crash are not intrinsic since the material, except for a few 2-3 nm wide bands which become amorphous, preserves integrity of its crystalline lattice. This rules out any explanation in terms of local melting, rebonding or influence of twinning and dislocation defects for the collapse. The same work, however, also showed that failure always occurs as these tiny amorphous bands, shock-induced in well defined lattice directions. More recently Ge $et\ al$. [5] have shown that similar structures can be produced also under quasi-static conditions, hence they are not an exclusive result of a kinetic process, and they contain free carbon, at least partly in graphitic, $sp^2$ hybridization.

Furthermore, even though the boron carbide microstructure has been widely accepted to be formed by 12-fold icosahedra and 3-fold chains, the actual location of the carbon sites in such microstructure is not yet clear. Duncan et al [6] proposed fully boron icosahedra and fully carbon chains, i.e. $B_{12}(CCC)$. Theoretical works have generally suggested that a phase formed by icosahedra containing one carbon atom at one polar site, i.e. $B_{11}C(CBC)$, is more favored [7]. However, all the simulations also found that the differences in ground state energies between the two polytypes, and many other ones, such as, for instance, a combination of $B_{10}C_2(CBC)$ and $B_{12}(CBC)$ polytypes are small, so that the coexistence of many polytypes must be assumed, as corroborated by some experiments [8-10]. Nevertheless all these calculations, though qualitatively significant on the strong sensitivity of the boron carbide microstructure to limited amount of disorder, bear a little practical meaning. Indeed, it is the Gibbs free energy at the formation temperatures in determining the coexistence of multiple phases and polytypes, and not the zero-point energy as directly provided by DFT calculations which is simply one of the components of the Gibbs free energy.

In order to model the catastrophic failure of boron carbide at high impact pressures our work has been divided in two parts. First, we performed $ab\text{-}initio$ density functional theory (DFT) simulations in order to consider the role of disorder in boron carbide and calculate the Gibbs free energies of a number of polytypes, showing them to coexist in the actual material. Next, we focused on the consequences of our disorder model to the dynamic mechanical properties. Specifically, we have investigated the stability of $B_4C$ upon transformation into icosahedral boron ($B_{12}$) and free carbon. We show that the observed failure at the $H_{EL}$ is consistent with the conversion of one of the coexisting polytypes, $B_{12}(CCC)$ into $B_{12}$ and a-C. The other polytypes experience similar transformation, if any, only upon much higher impact pressures.

THEORY

Kohm-Sham DFT simulations were carried out using the Gaussian03$^{TM}$ package, fitting the electronic structure of solids and molecules using basis sets of Gaussian functions. We used the PBE generalized gradient approximation for the exchange-correlation term, as proposed by J.P. Perdew $et\ al$. [11]. From the total energies per atom ($E_0$), we calculate the Helmholtz free energies per site $F_i$ for a number of i-polytypes of $B_{4-x}C_{1+x}$ :

$$F \approx E_0 + U_0 + F_{vib}(T) \qquad (1)$$

where $U_0$ is the zero-point energy, and $F_{vib}$ the temperature-dependent vibrational energy within the quasi-harmonic approximation. For both, we have used a Debye approximation of the vibrational density-of states. From that, the relevant quantity to test a polytype stability, the Gibbs free energy

$$G_i = F + P \cdot V_i, \qquad (2)$$

can be easily extracted. where $V_i$ is the unit cell volume divided per number of atoms in the unit cell. Note that, at room pressure. it is almost irrelevant to deal with G and $F_i$, while the $P \cdot V_i$ term becomes important at high pressure.

Unlike the previous theoretical works, we have used a multi-polytype approach, accounting for the coexistence of different polytypes in the same lattice within a disorder potential, expressing the local statistical fluctuation of the potential energy at an atomistic level, as a consequence of the disorder occurring at the preparation conditions of the material. This approach is a rigorous quantification of the empirical chemical laws generally known as Ostwald and Ostwald-Volmer rules [13] indicating that a multi-phase system (a) will pass through all the metastable phases before reaching the ground state, and (b) in the presence of several stable/metastable phases. the lesser dense phase is formed first.

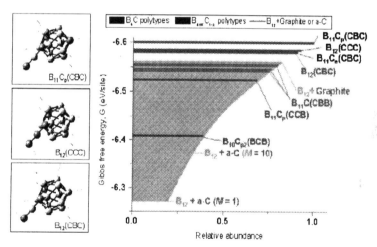

**Figure 1**. Some examples of Boron Carbide polytypes (left) and their relative abundance (right)

## MODELLING THE BEHAVIOR AT HIGH SHOCK PRESSURE

### Room pressure behavior

In the solid state systems, the Ostwald rules are generally difficult to quantify since phases "interact" in the sense that each one involves some lattice mismatching into others.

Therefore, the partition function of the system is very difficult to express. In boron carbide however, important simplification arises from the similarity in lattice constants of all the polytypes allowing them to be treated as a "non-interacting" system.

Under such a model, we get a number of metastable polytypes having Gibbs free energies lying within an energy range comparable to the kinetic energy $k_B T_s$ available at the synthesis temperature $T_s$. Of course, the synthesis temperature is neither instantaneously nor homogeneously reached. Thus, different regions of the material under synthesis experience energies E with a probability distribution

$$x_i(E) \sim \exp(-G_i/k_B T_s) \qquad (3)$$

and a number of polytypes, including those listed in **Fig.1**, are synthesized during growth. It has to be stressed that, under this approach, the Gibbs free energies per site, and not per unit cell, must be compared to the disorder potential. The same calculations, coupled with a Debye model and a Birch-Murnaghan equation of state are also used to extract the elastic moduli constants of the materials between room pressure and 40 GPa. They are found to be in agreement with those reported by Lee *et al.* [14] and experimentally measured by McClellan *et al* [15] with little dependence on the polytype.

Behavior under high pressures

Let us first observe that the connectivity of the icosahedral sites is very high, each one of them being connected with other 6 sites, 5 of them pertaining to the icosahedron. We then expect very high activation energy to extract an atom from the icosahedron. Our simulations show that the extraction energy is indeed higher for a polar site than for an equatorial one while only little activation energy is required to swap sites within a chain or an icosahedron. This sheds light on the transition path involved by the segregation of $B_{11}C(CBC)$ into separated phases of $B_{12}$ and graphitic C. In order to quantify the phenomenon, we can hypothesize that the transformation occurs in four steps, as in **Fig. 2**

(i)   Migration of the C site in the icosahedron from a polar to an equatorial site $[B_{11}C_{pl}(CBC) \rightarrow B_{11}C_e(CBC)]$;

(ii)  Migration of the B site position in the chain from the central to a boundary site $[B_{11}C_p(CBC) \rightarrow B_{11}C_e(BCC)]$ with the formation of an electronic defect state;

(iii) Swapping of the equatorial icosahedral C-atom with the boundary B-atom in the chain $[B_{11}C_e(BCC) \rightarrow B_{12}(CCC)]$ and finally,

(iv)  Coalescence of the as-obtained (CCC) chains on the (113) planes, through a rotation of their axis around the [001] vector. This eventually leads the (CCC) chains, already involving carbon planes, to finally relax into a hexagonal, graphitic lattice.

The energy minima between steps (i) and iv) correspond to the $B_4C$ polytypes in **Fig. 2**. Due to their similarity in force constants, each step can be considered separately and the activation energy of the process will correspond to that of the highest step, hence to the extraction energy of the $C_{eq}$ atom from the icosahedron (step iii). In order to be spontaneous, such an effect requires the icosahedron to become unstable. We stressed icosahedra of different composition at increasing pressures and we found the $B_{12}C_{eq}$ icosahedron, actually the less stable one, to be still stable at $P_1 = 40$ GPa.

We can therefore conclude that $B_{11}C(CBC)$, either with the C site in polar or equatorial position, is at least metastable up to 40 GPa, a value in agreement with the phase transition the transition pressure suggested by Voegler et al [3]. In contrast, the $B_{12}(CCC)$ polytype does not require any change in the icosahedron structure in order to transform into separated $B_{12}$ and graphitic phases. Its transition pressure $P_2$ will then correspond to the energy necessary to rotate the (CCC) chains along the [001] vector and align them on the (113) plane. This requires very small displacement of the boundary C-atoms in the chain, which may occur at much lower pressures than $P_0$.

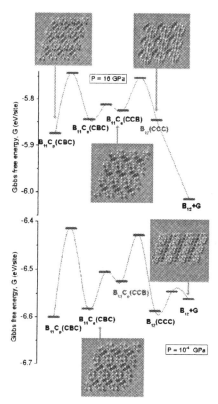

**Figure 2.** Sketch of the reaction steps i-iv) (see text) at room pressure (bottom) and 16 GPa (top)

Actually, when stressing $B_{12}(CCC)$ at increasing pressures we find that it becomes unstable at $P_2(\infty) = 6$ GPa. This pressure is just a slightly lower than the pressures $P(H_{F.L.}) \approx 7$-8 GPa occurring behind the shock wave at the failure conditions. Furthermore, the Gibbs free energy in finite aromatic carbon islands of M rings [16], is very little dependent on M

$$G(M) \sim 1.62 - 0.29 \cdot M^{-0.2} \qquad (4)$$

with little dependency on disorder or island amorphization [16]. Therefore G(M) it is just very slightly different in graphite (M→∞) and in graphitic amorphous carbon composed by distorted finite graphitic islands. Hence weak dependency of $P_2(M)$ on the degree of amorphization and disorder is expected, until an almost complete disappearance of the $B_{12}(CCC)$ polytype will occur, and the free carbon phase is easily produced due to shear induced amorphization during the impact.

Comparison with electron microscopy experiments

We have also investigated the most stable configuration of a segregated $B_{12}$- -free carbon phase, finding that it consists in carbon layers oriented along the (113) plane mixed with layers of $B_{12}$ icosahedra. This is in excellent agreement with the TEM images of Chen *et al* [4] (**Fig. 3**) and gives a full explanation of the relationship between the catastrophic failure of boron carbide and the appearance of an amorphous carbon amount as little as is the $B_{12}(CCC)$ concentration in the starting material. Appearance of bands of segregated free carbon and icosahedral $B_{12}$, having little elasticity and no plasticity, immediately terminates the elastic range of boron carbide. It prevents additional plastic range even if the residual non-segregated majority portion of the material would allow it.

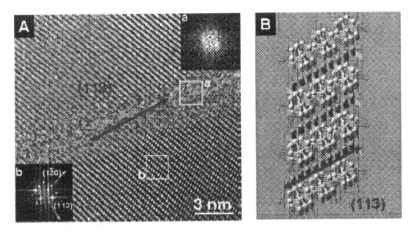

**Figure 3.** Comparison of **(A)** TEM image [4] and **(B)** simulated structure after shock.

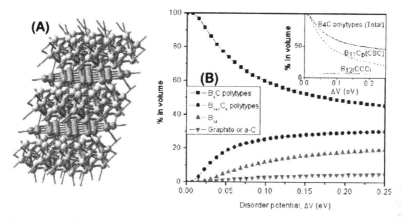

**Figure 4. (A)** Simulated periodical structure involving $B_{12}$ icosahedra and graphitic islands along the (113) plane of the pristine $B_4C$ lattice. **(B)** Relative abundance of the various polytypes and inclusions predicted as for eq. (3).

## PRESENCE OF INCLUSIONS

### Graphitic and amorphous Carbon

Graphite is often observed in boron carbides and even in $B_4C$ [17]. Our results show that besides the various polytypes, a phase involving separated boron and graphitic carbon can also be accommodated between the icosahedra without a significant loss in Gibbs free energy. The most favored arrangement of the segregated phases (average free energy per site $G_{B+G} = 6.560$ eV) involves graphitic layers along the intericosahedral voids corresponding to the reticular (113) plane and slightly displaced $B_{12}$ icosahedra, as shown in **Fig. 4A**. We can notice that the (113) plane is actually the same plane involved in the catastrophic failure as described in the previous paragraph.

The other arrangement of the segregated carbon phase is in the form of amorphous carbon (a-C) which can be present as a collection of graphitic islands of finite size, each one formed by M number of rings crosslinked by variable but generally a small amount of $sp^3$ carbon sites [16]. Segregation in the form of a-C is slightly less energetically favorable than graphite, as for eq. (4), but, the difference is still comparable to $\Delta V \approx k_B T_s$ at any value of M. Such considerations are quantified in **Fig. 4B** which describes the coexistence of the various boron carbide ($B_{4+x}C_{1-x}$) polytypes as a function of the disorder potential and the segregated boron / carbon phase. The figure shows that when increasing disorder at a constant stoichiometry (e.g. B/C = 4.0), the probability of understoichiometric polytypes (such as $B_{11}C_p(BCB)$ [18]) segregated boron icosahedra and graphite, or a-C, also increases.

Our results are consistent with recent TEM finding of graphite inclusions in commercial-grade boron carbide materials [19].

<u>Oxygen</u>

Fig. 5A reports the optimized structure of $B_{12}O_2$, which is still composed of $B_{12}$ icosahedra crosslinked by -O-O- chains. The spacing between the two oxygen sites ($\approx 3.0$ A) is much higher than in the $O_2$ molecule ($\approx 1.2$ A) pointing at the fact that O is also strongly bonded to the icosahedra. Nevertheless, the Gibbs free energy per site of $B_{12}O_2$ has the same values than in most the $B_4C$ polytypes, the differences being still within the disorder potential (**Fig 5B**).

If we distort the $B_{12}O_2$, unit cell in order to accommodate the -O-O- chains in the lattice of the $B_{11}C_p(CBC)$ polytype, the Gibbs free energy experiences just a very little decrease, passing from eV/site to eV/site. This means that, at least in principle, oxygen inclusions may easily coexist in the boron carbide lattice. Even if detailed calculations have still to be performed, the structure of $B_{12}O_2$ is more fragile at higher pressures, tending to transform into $B_{12}$ and $O_2$, with subsequent oxygen release, whereas the compression involves smaller intericosahedral distances for the -O-O- chains. Hence, oxygen incorporation might lead to brittleness in boron carbide.

However, we have to stress that while the Gibbs free energies are similar, the heat of formation is much higher in $B_{12}O_2$ than in $B_4C$ polytypes. This means that, under the current preparation conditions of boron carbide, it is unlikely that relevant amounts of bonded oxygen are incorporated in the material and, therefore, it is unlikely that the root causes of the failure of boron carbide can be due to oxygen-related brittleness. Furthermore, a failure related to oxygen release could not account for the formation of amorphous carbon bands in the material.

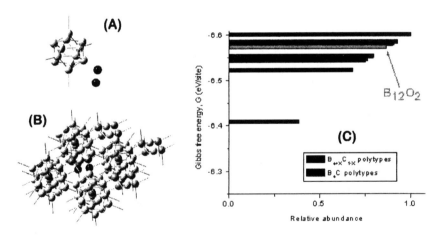

**Figure 5.** Structure of $B_6O$ **(A)** as a bulk material and **(B)** as an inclusion in a $B_4C$ lattice. **(C)** Gibbs free energy of $B_{12}O_2$ compared to that of the $B_4C$ polytypes

Silicon

Identifying the root causes of the dynamic failure of boron carbides opens up the possibility of tailoring their properties by synthesizing a new class of shock-resistant materials through a careful reduction of disorder during synthesis. However, as far as thermal disorder is unavoidable, such a route could be difficult to be applied to the actual, hot pressed, boron carbide materials. Therefore, in order to tackle such a problem, we studied the effects of silicon doping. **Fig. 6** reports the Gibbs free energy of the $B_{11}C_p$(CBC) and $B_{12}$(CCC) polytypes as a function of content of substituent silicon atoms [i.e. achieved by replacing a fraction of the pristine unit cells with $B_{11}Si_p$(CBC) and $B_{12}$(CSiC) respectively, which we find to the most stable $B_{12}SiC_2$ polytypes]. It can be observed that the difference in Gibbs free energy between the stable polytype, $B_{11}C_{1-y,p}Si_{y,p}$(CBC) and the most energetically favored minority polytype, $B_{12}$(CSi$_y$C$_{1-y}$C), strongly increases, by a factor 15, from $\Delta G = 0.015$ eV at 0 at % Si (see also **Fig. 1B**) to $\Delta G = 0.23$ eV at 6.7 at % Si. Hence, in the latter case, the concentration of the minority polytype is $\exp(-15) \approx 3 \cdot 10^{-7}$ times that occurring in absence of Si. In contrast the force on the atoms and their derivatives only experience little variations, corroborating the idea that the elastic constants do not experience large modifications at low enough Si doping.

It is important to observe that Si has been guessed to be the only hetheroatom leading to a reduction in concentration of the $B_{12}$(CCC) polytype in $B_4C$ [21]. Furthermore, in amorphous carbon, it is well known that even little amounts of Si strongly improve the sp$^3$ carbon content [22]. Especially, sp$^3$, diamond-like, carbon hybridization results in 3-D structures which would prevent segregation of carbon in 2-D amorphous bands as in **Fig. 3**. This suggests that Si doping could be a practical way to increase the activation energy for carbon segregation, avoiding the failure of boron carbide. The $H_{EL}$ of Si-doped boron carbide should be comparable, or even higher, to that of an intrinsic boron carbide with the $B_{12}$(CCC) polytype suppressed.

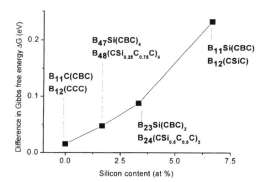

**Figure 6**. Gibbs free energy of the $B_{11}C_p$(CBC) and $B_{12}$(CCC) polytypes as a function of content of substituent silicon atoms . It can be observed that the difference in Gibbs free energy between the stable polytype, $B_{11}C_{1-y,p}Si_{y,p}$(CBC), and the most energetically favored minority polytype, $B_{12}$(CSi$_y$C$_{1-y}$C), increases with the silicon content. Hence silicon-doped boron carbide is less affected by disorder than the undoped one.

CONCLUSIONS

In summary, our results resolve the longstanding question of why boron carbide catastrophically fails just above the Hugoniot elastic limit and suggest an actual route to prevent such an effect. We have shown that the presence of small amounts of a minority polytype, $B_{12}(CCC)$, is responsible for the instability of the whole material at 6 GPa, close to the pressure just below the shock wave at the Hugoniot conditions. Therefore, the absence of plasticity is not an intrinsic property of this material, but simply of the tiny bands of amorphous carbon and boron icosahedra in to which $B_{12}(CCC)$ transforms. Furthermore, we have examined the presence of various inclusions in boron carbide, such as amorphous and graphitic carbon, oxygen and silicon.

We suggest silicon doping to be an actual route to tackle such an issue. Indeed, silicon-doped boron carbide involves larger differences within the Gibbs free energies of the various polytypes, resulting, at a constant disorder potential, in much higher fraction of the most stable polytype.

REFERENCES

[1] N.K. Bourne, Proc. R. Soc. London A 458 (2002) 1999
[2] D.S. Cronin, K. Bui, C. Kaufmann, G. McIntosh, T. Berstad, in 4th European LS-DYNA Users Conference Proceedings, DI47-57. Ulm, Germany, 2003
[3] T.J. Vogler, W.D. Reinhart, L.C. Chhabildas, J. Appl. Phys. **95** (2004) 4173.
[4] M. Chen, J.W. McCauley, K. Hemker, Science 299 (2003) 1563
[5] D. Ge, V. Domnich, T. Juliano, E.A. Stach, Y. Gogotsi, Acta Mater. **52** (2002) 3921
[6] T.M. Duncan, J. Am. Chem. Soc. **106** (1984) 2270
[7] D.M. Bylander and L. Kleinman, Phys. Rev. B**42** (1990) 1394
[8] D. Emin, Phys. Rev. B **38** (1988) 6041
[9] P. Lunca-Popa, **J I Brand, S. Balaz, L.G. Rosa, N.M. Boag, M. Bai, B.W. Robertson**, P.A. **Dowben**, J. Phys. D: Appl. Phys. 38 (2005) 1248
[10] Y. Feng, G. T. Seidler, J. O. Cross, A. T. Macrander, Phys Rev B**69** (2004) 125402.
[11] J.P. Perdew, K. Burke, M. Ernzerhof, Phys. Rev. Lett. 77 (1996) 3865
[12] Gaussian03, Revision C.02, M.J. Frisch *et al*, Gaussian Inc., Wallingford CT, 2004
[13] A. Barthl, Int J. Refr. Metals and Hard Materials, **14** (1996) 145
[14] S. Lee, D.M. Bylander, L.Kleinman, Phys Rev B **45** (1992) 3245
[15] K.J. McClellan, F. Chu, J.M. Roper, I. Shindo, J. Mater Sci **36** (2001) 3403
[16] J. Robertson, Adv. Phys. **35** (1986) 317; Diamond Related Mater. **4** (1995) 297.
[17] F.Thévenot, J Eur Ceram Soc **6** (1990) 205
[18] D.M. Bylander, L. Kleinman, S. Lee, Phys. Rev. B **43** (1991) 1487
[19] M.W. Chen, J.W. McCauley, J.C. LaSalvia, K.J. Hemker, J. Am. Ceram. Soc. **88** (2005) 1935
[20] S. Lee, S.W. Kim, D.M. Bylander, L. Kleinman, Phys. Rev. B **41** (1991) 3550
[21] H. Werheit, T. Au, R. Schmechel, S.O Shalamberidze, G.I Kalandadze, A.M. Eristavi, J Sol State Chem 154 (1999) 79
[22] C. De Martino, F. Demichelis, A. Tagliaferro, Diam. Related Mater., **6** (1997) 559

# ANALYSIS OF TEXTURE IN CONTROLLED SHEAR PROCESSED BORON CARBIDE

D. Maiorano, R. Haber, G. Fanchini
Rutgers University
607 Taylor Rd
Piscataway, NJ 08854

## ABSTRACT

Directionally induced texture through controlled shear processing of boron carbide is being investigated as a method of enhancing boron carbide's ballistic response. Texture was induced in boron carbide samples through hot pressing and tape casting. Texture analysis was conducted through optical microscopy techniques; Raman spectrometry studies were conducted on polished hot pressed samples. Texture was shown to be present in both tapes and hot pressed samples.

## INTRODUCTION

Intensive research is being conducted on developing new, superior materials for ceramic armor plates. One of the best materials, boron carbide, exhibits what appears to be an amorphous phase transition above the Hugoniot Elastic Limit (HEL) which Hemker et al have theorized may account for anomalous ballistic response at very high impact velocities imparting a ballistic impact of greater than 19 GPa. Despite much research into this amorphous phase, however, there has been no conclusive proof for an explanation of the origin or mechanism of formation of the amorphous phase.[1] It is believed, though, that this amorphous phase is caused in part by the substantial elastic modulus anisotropy of a full order of magnitude which exists in boron carbide crystals, shown in Figure 1, and contributes to fracture under ballistic testing due to increased intergranular fracture. Thus, it may be desirable to eliminate this amorphous phase transition, and induced texture is thought to be a possible convenient solution.

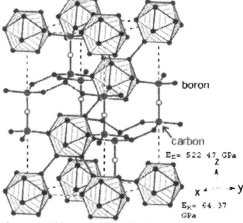

Figure 1. The boron carbide crystal with highly anisotropic elastic modulus.

By orienting the grains within a microstructure, the elastic mismatch between grains would be reduced, which is expected to increase the fracture toughness of the textured body by reducing localized stresses. This reduction in localized stress is predicted to be capable of preventing the amorphous transition, especially when stress is applied along specific planes of interest in the body. Haber and Nycz [2] have shown that controlled-shear processing can induce texture in slightly anisotropic alumina bodies, as the work of Seabaugh et al shows in Figure 2.[3] Any of a number of controlled-shear processes can be used to study the induced texture, with tape casting chosen for this study due to its ease of use and prevalence within industry, enabling scale-up.

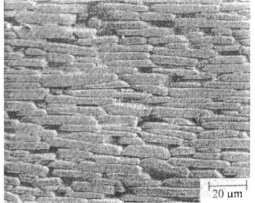

Figure 2. Example of a highly textured alumina microstructure. [3]

Similarly, hot pressing has also been shown to induce texture in bodies, as anisotropic grains orient perpendicular to the applied pressure during sintering. Boron carbide is routinely hot pressed, as it is one of the more common ways to produce bulk bodies. Therefore, hot pressing and tape casting of boron carbide was chosen to be examined in this study for the evidence of texture induced by each, with the ultimate goal of creating a hot pressed body composed of laminated tape cast parts.

There are several methods for determining texture within a body, mainly focusing on diffraction of various types. X-ray diffraction by the Lotgering Method, developed during F. K. Lotgering's work on ferrimagnetic oxides[4] is one of the most common, followed closely by use of x-ray diffraction rocking curves and neutron diffraction. It has also been shown that texture can be determined through the use of optical microscopy. When a highly anisotropic particle is studied under an optical microscope, the high degree of anisotropy will result in a degree of birefringence. This birefringence will cause a change in intensity of transmitted light when viewed with crossed polarizers, common on modern optical microscopes. Due to its relative ease of use and interpretation as well as high sensitivity to texture, the optical technique was chosen for a first study of texture in boron carbide samples.

Raman spectroscopy is often used to obtain information of the structure of the material being examined as Raman shifts can be exclusively attributed to specific vibrational modes. With the ability to study the material's response to a wide range of wavelengths of incident light with short test times, Raman is convenient as a quick and reliable method for determining many aspects of a sample.

EXPERIMENTAL

    10 μm powders of a platy morphology were obtained from UK Abrasives, Inc (Northbrook, IL), shown in Figure 3. Although boron carbide particles have a tendency to grow anisotropically, most commercially available powders have been milled to a fine spherical morphology. Obtaining the platy morphology powder was critical to this investigation, as spherical particles cannot exhibit preferential grain orientation. The powder was used for both tape casting and hot pressing studies.

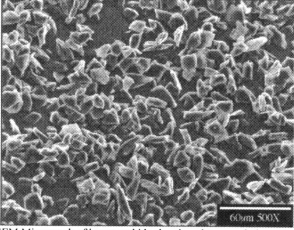

Figure 3. SEM Micrograph of boron carbide showing platy morphology (Courtesy UK Abrasives)

    Tape casting of boron carbide has not been extensively investigated in the past, and compositions determined by S. Tariolle et al proved to be incompatible with the current work.[5] Upon successive casts at varying compositions, an ideal working composition was determined of 70.5% by weight ceramic, 1.5% carbon sinter aid as Huntsman HX3 proprietary boron/carbon surfactant (H.C. Spinks, Paris, TN), 4.5% dispersant, and 23.5% binder/plasticizer, and was cast at 50% solids loading. The tapes were formulated by mixing the ceramic, sinter aid, and dispersant in deionized water on a ball mill for 24 hours. At this point the binder was added and the slip was milled for another hour before casting at 7.3 cm/s and a doctor blade gap of 250 μm. After casting, tapes were laminated into monoliths approximately 2.25 cm thick by pressing at 96 MPa in a Carver dry press. Binder removal was accomplished through heating of the laminated samples to 450 °C at 1 °C/min and held at temperature for half an hour.

    Hot pressed samples were produced by first producing green compacts of the boron carbide powder with 2% by weight Huntsman HX3 with 7% deionized water to improve flow and dry pressing to 96 MPa. After compaction, the samples were held at 110 °C for a minimum of 24 hours to remove the water. The compacts were then loaded into a Vacuum Industries inductively heated hot press, with boron nitride applied to the die and rams to prevent leaching of carbon into the samples from the graphite die. The press was then operated at 38 MPa with holds varying between 2000 °C to 2075 °C at lengths of 15 minutes to one hour to determine optimal conditions for achieving theoretical density.

Samples were then prepared for study through optical microscopy and Raman spectroscopy. For optical microscopy examination of texture, tapes had to be mounted on optical slides using thermoset resin and polished. As boron carbide scatters light quite strongly, samples were polished until very thin to enable study on the optical microscope. Top and bottom surfaces of the tapes were examined for a full range of cross polarization. Hot pressed samples were cut to expose faces parallel to the direction of hot pressing. Samples were then mounted in epoxy and polished to enable Raman spectroscopy upon faces parallel and perpendicular to the direction of hot pressing. Raman spectra were taken using an incident beam of 785 nm.

RESULTS AND DISCUSSION

When the tape cast samples were examined for induced texture, the optical technique proved useful for showing that there was indeed texture. Figure 4 shows the comparison of transmitted light through the top surface of the tapes, with a noticeable change in transmitted light upon cross polarization. These intensities help to prove that tape casting has indeed induced texture in the samples. Due to the high degree of scatter through boron carbide and the degree of polishing required to obtain a thin sample from the tapes, now that texture has been established as occurring in the tapes x-ray diffraction techniques can be employed to more quickly obtain data on the texture of the samples.

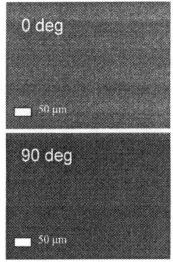

Figure 4. Optical micrographs of tape cast samples showing variance in transmitted light upon polarization of incident light.

High density samples were desired from the hot press runs for study under Raman spectroscopy. Density was taken from samples pressed at 38 MPa and 2000 °C, 2050 °C, and 2075 °C. Runs at 2000 °C resulted in pieces of 90% theoretical density. The samples pressed at

2050 °C resulted in densities of approximately 96% theoretical, and those pressed at 2075 were 98% of theoretical density. In addition, the samples pressed at 2075 °C were studied for the effect of hold time at temperature. Two samples were produced at 2075 °C, one with a hold time of 15 minutes and the other with a hold time of one hour. When density measurements were taken, the sample held at temperature for one hour had a slightly higher density, but when examined under optical microscopy after polishing, exhibited a much greater degree of surface porosity compared to the sample held at temperature for 15 minutes. This difference in porosity was attributed to the presence of free graphite in the sample, present through the excess carbon in the original working powder and the additive Huntsman surfactant. This excess graphite would have been pulled out from the sample during grinding and polishing, resulting in the porosity seen when optically examined. Hardness data was taken on these samples for comparison to conventional hot pressed fully dense samples and was adversely affected by the porosity. It was difficult to determine indent size due to the great degree of porosity present upon polished sections from both surface and interior faces of the sample, and upon Knoop testing at loads of 0.1 kg, the samples exhibited hardness values two orders of magnitude below values normally experienced in dense parts.

Figure 5 displays the Raman spectra intensity patterns obtained using a 785 nm incident beam upon hot pressed samples. These patterns depict a fairly clean boron carbide pattern, with disorder peaks around 300 cm$^{-1}$, which have yet to be fully identified as to the cause of disorder and what information can be obtained from the disorder peaks. The Raman spectra shows a small but noticeable amount of free carbon present in the sample, with D and G peaks for carbon present at 1300 and 1600 cm$^{-1}$, respectively, as shown in Figure 5c. The relative intensities of the D and G peaks are slightly stronger when the sample is examined perpendicular to the direction of hot pressing than when examined parallel to the hot pressing direction. As graphite tends to have a stronger Raman signal when vibrating aromatically in plane, the stronger relative intensities show that the excess carbon in the sample has experienced some alignment perpendicular to the direction of hot pressing. Similarly, the boron carbide spectra itself is slightly stronger perpendicular to the direction of hot pressing over parallel to the direction of hot pressing, showing alignment of grains of the boron carbide merely from the press. This is encouraging as it can be used to predict hot pressing of tape cast samples will reinforce the texture induced through the shear of casting.

CONCLUSION

Texture was shown to be present in both tape cast and hot pressed samples of platy boron carbide. Optical microscopy showed a change in intensity of transmitted light upon cross polarization indicative of a textured anisotropic microstructure. Hot pressed samples were evaluated to determine optimal pressing protocols to approach theoretical density. Samples were shown through hardness testing and optical microscopy to have a great deal of porosity. Porosity was caused by excess carbon present in the sample, shown through Raman spectroscopy to have been aligned perpendicular to the direction of hot pressing. Future work will determine the effect this induced texture has upon the mechanical and ballistic properties of bulk boron carbide samples.

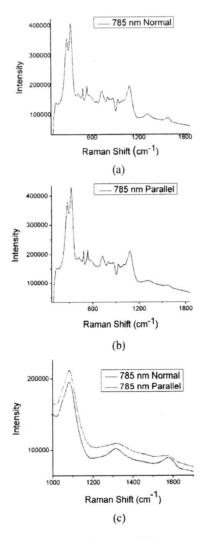

Figure 5. Raman spectra for hot pressed boron carbide (a) perpendicular to the direction of hot pressing, (b) parallel to the direction of hot pressing, and (c) the free carbon peaks compared between directions of study.

REFERENCES

[1]M. Chen, J. McCauley, K. Hemker, Science 299 (2003) 1563

[2]S. Nycz, R. Haber, "Controlling Microstructural Anisotropy During Forming", Ceram. Eng. and Sci. Proc., **26** [8] 45-52 (2005).

[3]Seabaugh, Kerscht, Messing. "Texture Development by Templated Grain Growth in Liquid-Phase-Sintered a-Alumina." J. Am. Ceram. Soc., **80** [5] 1181-1188 (1997).

[4]F. K. Lotgering, J. Inorg. Nucl. Chem [9] 113-123 (1959) .

[5]S. Tariolle, C. Reynaud, F. Thevenot, T. Chartier, J.L. Besson, "Preparation, microstructure and mechanical properties of SiC-SiC and $B_4C$-$B_4C$ laminates." J. Solid State Chem., [177] 487-492 (2004).

# Damage and Testing

# PROGRESS IN THE NONDESTRUCTIVE ANALYSIS OF IMPACT DAMAGE IN TiB$_2$ ARMOR CERAMICS

Joseph M. Wells
JMW Associates
102 Pine Hill Blvd, Mashpee, MA 02649-2869
(508) 477-5764   jmwconsultant@comcast.net

## ABSTRACT

The main objective of this paper is to provide an updated overview of the recent and more interesting damage characterization results revealed from a continuing non-destructive examination of encapsulated TiB$_2$ ceramic targets impacted with 32gm high velocity projectiles with an L/D ratio of 20. X-ray computed tomography, XCT, and advanced voxel analysis and visualization software (Volume Graphics StudioMax v1.2.1) techniques are utilized to provide unprecedented diagnostic flexibility into the volumetric characterization and analysis of complex in situ ballistic impact damage. These techniques facilitate the creation of digitally rendered 3D solid object reconstructions, arbitrary virtual planar sectioning, variable transparency and segmentation of both projectile fragment and cracking damage morphology and distribution, virtual metrology and 3D visualization of damage features of interest, and impact induced porosity analysis. Examples of several of these unique NDE damage observations are provided and discussed for improved appreciation, understanding, and cognitive visualization of the complex ballistic impact damage occurring in impacted TiB$_2$ armor ceramic targets.

## INTRODUCTION

To better design, develop and evaluate lighter and more efficient ceramic armor materials, new conceptual material/target configuration modeling approaches, damage diagnostic and analytical techniques, and interdisciplinary collaborations are needed. Their purpose would be to identify, diagnose, analyze, assess, and ultimately control (suppress, mitigate, or diffuse) the extent and effects of various forms of impact damage on the ballistic performance of armor ceramics. Thus, it is essential to develop an improved methodology for characterization, visualization and understanding of the actual ceramic impact damage that occurs under various ballistic impact conditions.

## BACKGROUND

Impact damage of highly constrained or encapsulated armor ceramics consists essentially of micro- and meso-scale features on both the surface and in the interior of the target material. To date, it has been impractical to study the interior armor ceramic target damage details in real time during the impact event. Post-impact induced cracking has frequently been observed to have various presented forms such as radial cracks, ring or circular cracks, conical cracks, and laminar or lateral cracks. An excellent example of these "traditional" forms of mesocracking observed on 2D ceramographic planar section recently reported by Lasalvia[1] is shown in Figure 1. In addition, after penetration has occurred, high density projectile fragments are frequently located imbedded deeply within the impacted ceramic target. Furthermore, additional indications of "nontraditional" damage features (including spiral & hourglass-shaped ring cracking morphologies, non-uniform sub-surface radial expansion, raised impact surface steps, impact induced porosity, etc.) have been observed and reported by Wells et al[2-11]. It should be noted that all such damage features are intrinsically 3-dimensional in nature and/or distribution and preferably should be analyzed as such.

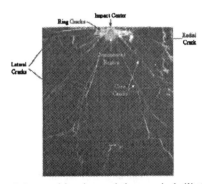

Figure 1. Traditional forms of impact cracking damage revealed by the destructive examination of a SiC$_N$ armor ceramic by J.C. LaSalvia et al.[1]

Because the ballistic impact event and the resulting internal damage in ballistic targets are essentially 3-dimensional phenomena, the author advocates the utilization of a non-intrusive examination technique with the capability of revealing the true 3D damage morphological details at desired locations throughout the entire volume of the target material. Such a non-intrusive NDE modality is uniquely achieved with industrial X-ray Computed Tomography, XCT. The XCT approach allows the complete volumetric digitization (density mapping) of the target sample, which is then subsequently diagnosed mainly by advanced voxel analysis and visualization software techniques. The essence of XCT is that of using an x-ray beam to penetrate a rotating object and capturing the transmitted x-ray energy by an electronic detector array. Sophisticated detector signal processing enables the triangulated "density" mapping of the object volume as a contiguous series of thin (~0.5mm) axial slice images. Once the XCT digitized scan file is imported into an advanced 3D voxel analysis and visualization software package, the fully digitized virtual solid object can be reconstructed in 3D and various damage features are clearly discernable from the base ceramic. The specific software prominently utilized in this work is Volume Graphics StudioMax, v1.2.1, although other similar software is also commercially available. Amongst the multiply capabilities this software provides are:

- Sophisticated image analysis and visualization capability to process, analyze and visualize voxel/volume data.
- Up to 3 GB of memory utilization with Windows XP Professional OS
- Multiple Import/Export File Formats
- Virtual Metrology Capabilities
- Variable Transparency & Virtual Sectioning
- Surface Extraction
- Segmentation
- Porosity/Defect Analysis
- Wall Thickness Analysis
- Stereo Viewing Tool

The present paper reports on updated impact damage observations and characterization results in a TiB$_2$ armor ceramic. Results obtained using this software and various diagnostic techniques include: 3D solid object reconstruction, impact surface topological examination, virtual sectioning, cracking damage segmentation, residual projectile fragment segmentation, statistical analysis of impact induced porosity, and the virtual metrology and cognitive visualization of various damage features of interest.

## DAMAGE CHARACTERIZATION OF IMPACTED TIB$_2$ ARMOR CERAMIC TARGETS
Target Samples

Three TiB$_2$ target ceramic disks, measuring ~72 mm in diameter by 25 mm in thickness, were each encapsulated in a welded case of Ti-6Al-4V alloy. The first sample, designated S1wo, did not have a 17-4 PH steel ring shrunk fit on its outer diameter before encapsulation as did the remaining two. The shrunk fit ring provided a compressive pre-stress on the second and third TiB$_2$ samples, designated S1w and S2w respectively, prior to their encapsulation. Following impact by a high velocity sub-scale 32 gm tungsten alloy projectile (L/D=20), it was necessary to remove each TiB$_2$ ceramic target from the outer 15 x 15 x 6.4 cm encapsulation package prior to XCT examination due to the limited penetration capability of the BIR 420kv x-ray facility utilized at ARL.

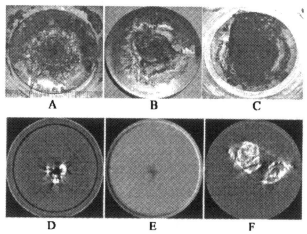

Figure 2. Macro-photographs (A, B, C) and XCT mid-thickness XCT axial scans images (D,E, F) of TiB$_2$ armor ceramic disk samples S1wo, S1w and S2w respectively.

Figures 2A through 2C reveal macro-photographs of the impacted exterior surface of the three target ceramic disks. Interior 2D axial XCT scan images at the approximate mid-thickness of each of the three target disks are shown in figures 2D through 2F respectively. Residual tungsten alloy projectile fragments (white) and internal damage features (dark) are prominently visible in figures 2D and 2F, while there is a complete absence of fragments and only very faint cracking damage features observed in the center of figure 2E. Thus the addition of the 17-4 PH steel compression ring significantly reduced the penetration and the damage level in sample S1w over that of sample S1wo. The target S2w also had a compressive ring and was able to sustain two individual and sequential impacts in this configuration. Note the two distinct mid-thickness agglomerated tungsten fragments visible in figure 2F.

Impact Surface Topological Observations

Two 3D solid object reconstructions for target sample TiB$_2$ S1wo are shown in figure 3. Three distinct raised circular steps surrounding the central impact crater are observed on the impact surface. Multiple radial cracks transverse to these surface steps are also observed on the impact surface, some of which appear not to extend to the same point in the central cavity. The inner most circular step was determined to be the thickest and the outer most step the thinnest, although the thickness of each step was somewhat variable along its circumference.

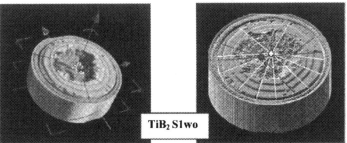

Figure 3. Reconstructed 3D solid object images show impact surface topological features of TiB$_2$ S1wo target sample. Note not all of the surface radial crack extensions intersect at the same locus.

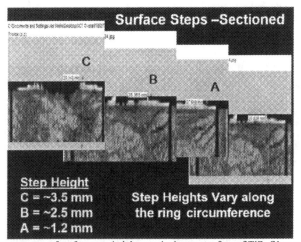

Figure 4. Measurements of surface step heights on the impact surface of TiB$_2$ S1wo sample on various through-thickness virtually sectioned planes.

Surface step height observations of the three concentric surface rings are indicated in figure 4. The lighter gray values of the impact surface rings in these images are quite distinct and discernable from the darker gray values of the ceramic itself. The gray values of these surface steps are also somewhat darker than the gray levels of the major projectile fragment shown in subsequent figures. These differences are indicative of the significantly higher density of these surface steps relative to the TiB$_2$ ceramic (but still lower density than the major fragment) and strongly suggest an outward radial surface flow of "semi-fluid" material with substantial tungsten alloy projectile content mixed with ceramic cavity debris along the impact surface.

Near Surface Radial Expansion Non-uniformity

Indications of non-uniform surface expansion are observed to within a depth of about 5 mm from the impact surface on target sample TiB2 S1wo as shown in figure 5. The largest indication of radial expansion was measured along diameter "A" as ~0.55 mm when comparing axial scan images #51 and #41. Smaller values of radial expansion were measured along diameters "B, C, & D". The nominal depth of this radial expansion is very similar to the depth of the radial cracks observed originating on the impact surface and propagating down to a measured depth of ~5.2 mm.

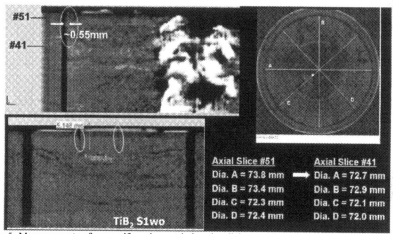

| Axial Slice #51 | Axial Slice #41 |
|---|---|
| Dia. A = 73.8 mm ⟶ | Dia. A = 72.7 mm |
| Dia. B = 73.4 mm | Dia. B = 72.9 mm |
| Dia. C = 72.3 mm | Dia. C = 72.1 mm |
| Dia. D = 72.4 mm | Dia. D = 72.0 mm |

Figure 5. Measurements of non-uniform impact induced radial expansion along diameters A, B, C, & D shown on axial slice #51 near the impact surface of sample target TiB$_2$ S1wo. Note similar depth of radial cracks (~5.2 mm) to depth of radial expansion depth (~5 mm).

Residual Projectile Fragments

The greater bulk of the residual projectile fragments reside in the interior of the penetrated ceramic target as indicated in figure 6, created with the virtual opacity of the ceramic target being gradually reduced until only the higher density fragment components of the XCT data are observed (C). This figure includes reconstructed semi- and fully transparent 3D solid object images including a virtually sectioned half disk (A) and full disk images (B&C). The high density tungsten alloy projectile fragments appear here in white localized near the center of each image and extending through the target ceramic thickness. Also visible (in A & B) are the three somewhat lower density mixed rubble surface steps on the observable top impact surface. In the case of the double sequential projectile impact on target S2w, the fully opaque 3D solid object reconstructed image is shown in figure 7(left)., while the fully transparent view of the high density dual (agglomerated) projectile fragments contained within are isolated and displayed in figure 7(right).

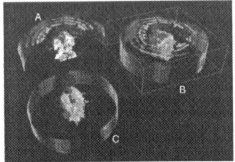

Figure 6. Reconstructed transparent voxel images of the 3D solid object TiB$_2$ S1wo disk target are shown as half virtual section (A) and full (B&C) renderings revealing localized internal residual projectile fragments (white areas).

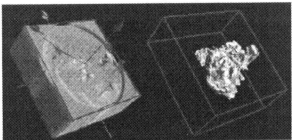

Figure 7. Shows fully opaque image of TiB$_2$ S2w (left), and the corresponding fully transparent image isolating the consolidated projectile fragments (right).

Previously, the discrimination and isolation of impact damage features from the visually obscuring opacity of the bulk ceramic was accomplished with the use of virtual point cloud images. Essentially, point clouds are constructed by using only selective threshold gray values of the feature(s) of interest while excluding all other non-related XCT data in a subsequent image reconstruction. Point cloud images of the respective residual *projectile fragments* of TiB$_2$ targets S1wo & S2w are shown in figure 8. These point clouds were developed early on in the initial

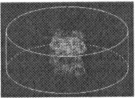

Figure 8. Earlier point cloud images of residual projectile fragments in TiB$_2$ S1wo ceramic target (left) and TiB2 S2w ceramic target (right).

characterization efforts and may be compared with the more recently developed transparent images of figures 6 & 7 above, where the same projectile fragments are segmented from the surrounding bulk ceramic and are viewed directly in considerably greater clarity and detail.

Interior Ring Cracking

Earlier point cloud images of the meso-scale impact *cracking* damage are shown in figure 9. The toroidal-like volume (left) of the interior cracking damage observed through the target sample thickness is apparent as well as the narrowing diameter at the sample mid-thickness (hourglass and spiral cracking damage features-right).

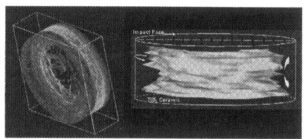

Figure 9. Modified images of isolated internal cracking damage in TiB$_2$ sample S1wo (left) and the hourglass-shaped edge contour (arrow) and the spiral cracking indications on the surfaced point cloud half section (right).

Additional indications of the hourglass cracking damage morphology are observed in the sequential viewing of planar XCT axial scans of target TiB$_2$ S1wo. Concentric circular ring cracks observed in the axial slice images (see figure 10, left) were found to decrease in diameter from the impact face to the mid-thickness of the sample and then increase in diameter as the back face of the sample is approached. Such observations substantiate the hourglass profile detail of figure 9. and also suggest a series of concentric "hourglass-shaped" through thickness ring cracking morphologies as shown schematically in figure 10. Additional indications of spiral cracking impact damage have been observed in TiC ceramic[10] and monolithic Ti-6Al-4V metallic targets[11] as well.

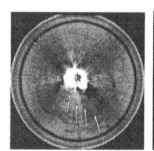

Figure 10. Concentric ring cracks (shown at arrows) in axial slice image of TiB$_2$ S1wo target (left) and 3D schematic of concentric ring cracks forming the through thickness hourglass morphology (right).

Impact Induced Porosity Observations

Another damage characterization manifestation in impacted ceramics recently explored is that of impact-induced internal voids or porosity. Using the defect analysis tool of the VGStudioMax software, completely enclosed multiple small pores with a volume up to 2.30 mm$^3$ were detected and categorized by volume as shown in figure 11. Any pore not completely enclosed by having virtual access to the surrounding exterior air environment was excluded from this analysis. The porosity observed is most abundant at the smaller pore volumes ($< \sim 1.2$ mm$^3$) and appears to be distributed throughout the ceramic sample volume with some preferred localization along internal damage cracking features. The aspect ratio of most pores was found to be asymmetric indicating an appreciable deviation from general sphericity. A total of 4392 pores in this volume range were detected with a cumulative pore volume of 2006 mm$^3$. For the target sample volume of 103,363 mm$^3$, this calculates to a total porosity level of $\sim 1.94\%$.

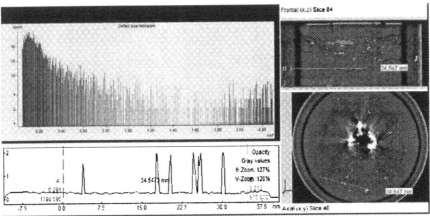

Figure 11. Example images of porosity analysis in TiB2 sample S1wo. Histogram of porosity frequency versus pore size in mm$^3$ (upper left), frontal slice (upper right) and axial slice (lower right) indicating impact induced internal porosity and profile trace (lower left) of porosity located along the fiducial line shown.

3-D Cracking Damage Quantification and Mapping

Efforts[7] at quantifying the magnitude of the meso-scale cracking damage as a function of sample radius and sample depth produced results as shown in the 3D plots of figure 12. The methodology utilized[7] was a relatively straight forward, but manually tiresome, counting of the damage voxels as a function of radius on each successive axial XCT slice. The ratio of the damage voxels to the total voxels contained in the local area being analyzed provides a value for the damage fraction at that radius. The angular orientation of individual damage voxels (i.e. theta value at a given radius and slice depth) is a more complex consideration and was not included in this data; thus the data presented should not be assumed to be axisymmetrical. Consideration is being given to automating this quantitative damage mapping method in the future. The process is more difficult, however, with the presence of the highly x-ray absorbing (high density) residual projectile fragments which introduces considerable noise (artifacts). Consequently, a method of electronically filtering the projectile fragment from the data was employed with the results compared in figure 12. More damage at the lower radii was captured without the obfuscation of the projectile fragments being included in the quantitative analysis as shown in the left hand side of figure 12.

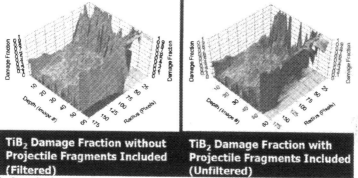

**TiB₂ Damage Fraction without Projectile Fragments Included (Filtered)**   **TiB₂ Damage Fraction with Projectile Fragments Included (Unfiltered)**

Figure 12. Quantitative 3D plot of damage fraction in TiB₂ S1wo sample without (left) & with (right) the penetrator fragments included in the quantitative asymmetrical damage analysis. Filtering of the penetrator fragments reveals more cracking damage.

## POTENTIAL IMPACT OF XCT DAMAGE CHARACTERIZATION

The evolution of the multiple meso-cracking damage features observed via XCT and apparently leading to the gradual loss of intrinsic structural constraint resisting the outward expansion of the comminuted ceramic was previously suggested by the author[5, 9] to advance the concept of the role of impact damage influencing the delay and resistance to penetration. A rough schematic of the impact event is shown in figure 13 illustrating observations from experimental "interface defeat" experiments[12] that despite the occurrence of substantial meso-scale cracking, the damaged ceramic matrix material retains adequate structural integrity to resist

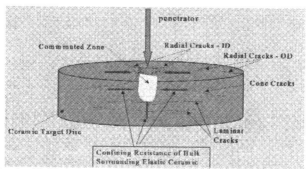

Figure 13. Sketch of ballistic impact event shows schematic constraint and confinement of the central comminuted zone by the surrounding meso-scaled damaged bulk ceramic.

the radial expansion of the comminuted zone, thus preventing penetration. At increased projectile velocities, the extent of meso-cracking damage increases with a concurrent decrease in the bulk ceramic structural integrity until the point is reached where the comminuted zone immediately below the impact location can no longer be constrained and penetration commences.

A recent study of the effect of increasing impact velocity (50-500 m/s), using WC-6Co spheres of 6.35 mm diameter, on the creation and evolution of interior impact damage in SiC & TiB2 armor ceramic targets (25.4 mm dia. x 25.4 mm length) has been published by LaSalvia et al[1]. However, these authors, to date, have used only physical sectioning, ceramographic preparation and 2D planar damage observations to characterize the details of the impact meso-scale damage. Currently this author and his collaborators at ARL[14] are using XCT diagnostics to explore the volumetric impact damage in SiC$_N$ armor ceramic targets of similar size to develop a non-intrusive characterization and analysis of the 3D impact damage in this material. Realistically, such impact damage characterizations should prove useful in furthering our collective understanding of the complex details and morphologies of ballistic damage features and their relationships to both penetration mitigation and the overall ballistic performance of armor ceramics.

SUMMARY

The availability of various nondestructive XCT diagnostic techniques now capable of examining the volumetric impact damage features in armor ceramics is considered a significant step forward in our evolving capability to characterize such damage. Volumetric characterization is considered important since the impact event and the resulting damage in real ceramic materials are essentially 3D phenomena. Several results to date indicate considerable 3D asymmetry in the morphologies of the various damage modalities, which may remain undetected in a 2D analysis.

Unfortunately, it is not possible at present to conduct high resolution XCT impact damage diagnostics in real time, but rather we are limited to non-intrusive post-impact investigation. While considerable progress has been achieved in the volumetric characterization, visualization and analysis of ballistic impact damage through the use of x-ray computed tomography, it may yet be premature to accurately assess the long term potential impact of the relatively recent introduction of this technology.

The XCT diagnostic capabilities presented above are currently considered quite useful in furthering our collective understanding of the details and complexities of volumetric impact damage in the case of TiB$_2$. To date, relatively few other armor ceramic impacted targets have been investigated with this XCT technology and caution must be exercised in making extrapolations of these results to other armor ceramic/ballistic impact conditions not yet actually investigated. Perhaps, assisted with these demonstrated XCT damage diagnostic capabilities, the focus of future ballistic studies will improve the integration of the role of damage, as well as penetration, within our knowledge base of what factors control the overall performance of armor ceramics. Still further improvements and refinements in the diagnostic and analytical capabilities of the XCT approach are realistically anticipated with an expanded experience base. Hopefully, both the experimental and the analytical modeling communities will become increasingly more familiar and interactively collaborative with future XCT studies of ballistic impact in armor ceramics.

ACKNOWLEDGEMENT

Acknowledgements are gratefully extended to N.L. Rupert, W.H. Green, J.R. Wheeler, and H.A. Miller at ARL for their technical contributions during the course of the original XCT examinations and damage analysis of the TiB$_2$ ceramic damage.

# REFERENCES

1.  J.C. LaSalvia, M.J. Normandea, H.T. Miller, and D.E. McKinzie, "Sphere Impact Induced Damage in Ceramics: I. Armor-Grade SiC and TiB2," Ceramic Engineering & Science Proceedings, v26, [7], 183-192 (2005).
2.  W. H Green, K.J. Doherty, N. L. Rupert, and J.M. Wells, "Damage Assessment in TiB2 Ceramic Armor Targets; Part I - X-ray CT and SEM Analyses", Proc. MSMS2001, Wollongong, NSW, Australia, 130-136 (2001).
3.  N.L. Rupert, W.H. Green, K.J. Doherty, and J.M. Wells, "Damage Assessment in TiB2 Ceramic Armor Targets; Part II - Radial Cracking", Proc. MSMS2001, Wollongong, NSW, Australia, (2001), 137-143.
4.  J. M. Wells, N. L. Rupert, and W. H. Green, "Visualization of Ballistic Damage in Encapsulated Ceramic Armor Targets", 20th Intn'l Symposium on Ballistics, Orlando, FL, ADPA, Vol. 1, (2002), 729-738.
5.  J. M. Wells, N. L. Rupert, and W. H. Green, "Progress in the 3-D Visualization of Interior Ballistic Damage in Armor Ceramics", Ceramic Armor Materials By Design, ed. J.W. McCauley et. al, Ceramic Transactions, v. 134, ACERS, 441-448 (2002).
6.  John M Winter, Jr., William J. Bruchey, Joseph .M. Wells, and Nevin L. Rupert "Review of Available Models For 3-D Impact Damage and The Potential For Incorporation of XCT Results", 21st International Symposium on Ballistics, Adelaide, Au , ADPA, v1, 111-117(2004).
7.  H.T. Miller, W.H. Green, N. L. Rupert, and J.M. Wells, "Quantitative Evaluation of Damage and Residual Penetrator Material in Impacted TiB$_2$ Targets Using X-Ray Computed Tomography", 21st International Symposium on Ballistics, Adelaide, Au , ADPA, v1, 153-159 (2004).
8.  Joseph .M. Wells, "On Non-Destructive Evaluation Techniques for Ballistic Impact Damage in Armor Ceramics", Ceramic Engineering & Science Proceedings, v26, [7], 239-248 (2005).
9.  Joseph .M. Wells, "Considerations on Incorporating XCT into Predictive Modeling of Impact Damage in Armor Ceramics", Ceramic Engineering & Science Proceedings, v26, [7], 51-58 (2005).
10. J.M. Wells, W.H. Green, and N.L. Rupert, "Nondestructive 3-D Visualization of Ballistic Impact Damage in a TiC Ceramic Target Material", Proceedings MSMS2001, 2nd International Conference on Mechanics of Structures, University of Wollongong, NSW, Australia, 159-165 (2001).
11. J.M. Wells, W.H. Green, N.L. Rupert, J. R. Wheeler, S.J. Cimpoeru, and A.V. Zibarov, "Ballistic Damage Visualization & Quantification in Monolithic Ti-6Al-4V with X-ray Computed Tomography", 21st International Symposium on Ballistics, DSTO, Adelaide, Australia, ADPA, v1, 125-131 (2004).
12. G.E. Hauver, P.H. Netherwood, R.F. Benck, and L.J. Kecskes, "Ballistic Performance of Ceramic Targets", Army Symposium on Solid Mechanics, USA, (1993).
13. J.C. LaSalvia, M.J. Normandea, D.E. McKinzie, and H.T. Miller, "Sphere Impact Induced Damage in Ceramics: III. Analysis," Ceramic Engineering & Science Proceedings, v26, [7], 193-202 (2005).
14. J.M. Wells, N.L. Rupert, D.E. McKenzie, & W.H. Green ,"XCT Observations of Impact Damage in SiC$_N$ Ceramic Targets," Presentation at 30[th] Intn'l Conf. on Adv. Ceramics & Composites, Cocoa Beach, FL, Jan. 2006.

ELASTIC PROPERTY DETERMINATION OF WC SPHERES AND ESTIMATION OF COMPRESSIVE LOADS AND IMPACT VELOCITIES THAT INITIATE THEIR YIELDING AND CRACKING[1]

A. A. Wereszczak
Ceramic Science and Technology Group
Metals and Ceramics Division
Oak Ridge National Laboratory
Oak Ridge, TN 37831-6068

ABSTRACT

Resonant Ultrasound Spectroscopy (RUS) was used to determine the elastic constants E, G, and ν of 6.35 mm diameter spheres of high purity WC (Roc500), WC-6%Co, and WC-12%Co. Two of the three elastic constants were independently determined using RUS; consequently, the determination of the third elastic constant proceeded without needing to assume a value for any one of the three. Secondly, 500g Vickers hardness was measured with each of the three compositions, and compressive loads and impact velocities to initiate yielding and cracking in the spheres were estimated from the combination of finite element analysis, the elastic properties measured from the RUS analysis, the measured hardnesses, and assumed tensile strengths for the WC compositions.

I. INTRODUCTION

Resonant ultrasound spectroscopy (RUS) is based on the principle that the mechanical resonance of a component depends on its shape, density, and elastic properties in a manner such that a measurement of the resonant frequencies provides a signature that is unique to the component. A resonance spectrum is generated by sweeping the frequency of an ultrasound signal applied to the component and by detecting its resulting resonance frequencies. Changes in the shape, density or elastic properties, or introduction of defects, lead to a variation of this signature. Advantages of RUS include: high sensitivity - very small deviations in E, G, and ν are detectable; quick testing - testing duration is only a few seconds, and; flexural, torsional, and longitudinal resonance modes are excited - this enables modal signature analysis that produces independent determinations of G and E or ν. The excitation of only one of those three modes is limiting because it requires the assumption of a value of E (or ν more commonly) to determine the values of all three elastic constants.

[1] Research sponsored by WFO sponsor US Army Tank-Automotive Research, Development and Engineering Center under contract DE-AC05-00OR22725 with UT-Battelle, LLC.

211

An example of a RUS spectrum is shown in Fig. 1. The frequencies at which there are resonances are unique to the geometry and density of the material and the specific values of E, G, and ν; knowledge of the geometry (i.e., accurate diameter measurement of a sphere, etc.) and density (a parameter easy to accurately measure with simple geometries) facilitate RUS analysis. This fact allows the RUS user to accurately determine E, G, and ν by numerically varying the values of E, G, and ν until a unique, three-way combination produces a match of experiment and modal resonance theory.

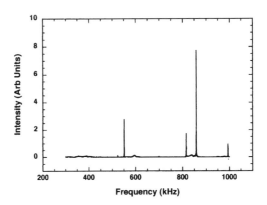

Figure 1. Example of a RUS-generated resonance spectrum.

The RUS's software does not accommodate analysis of materials in a spherical geometry so modal analysis resulting from finite element analysis, FEA, (e.g., ANSYS), in combination with the capturing of resonance frequencies from any 2 of the three resonance modes (i.e., flexural, torsional, and longitudinal), was used to accomplish this. Such a method has been used to determine the three elastic properties of ceramic armor tiles with a GindoSonic tester [1]. The method consists of (1) running three FEA modal analyses for a given geometry and material density, with the "bracketing" of two of the three elastic constants (e.g., E and ν), and (2) fitting their functionalities to the measured resonance frequencies. For example, if the ceramic was a WC-6%Co, and its E and ν were estimated to be 640 GPa and 0.20, respectively, then the three E - ν combinations for the three modal analyses could be chosen to be: 600 GPa - 0.15; 750 GPa - 0.15, and; 750 GPa - 0.25. These FEA results then enable the determination of the torsional frequency ($f_{tors}$) and a flexural resonant frequency ($f_{flex}$) as multilinear functions of E and ν for the modeled geometry and density; namely,

$$f_{tors} = A_t + B_t \bullet E + C_t \bullet v \quad \text{and} \tag{1}$$

$$f_{flex} = A_f + B_f \bullet E + C_f \bullet v \tag{2}$$

where $A_t$, $B_t$, and $C_t$ are regression constants for $f_{tors}$ and $A_f$, $B_f$, and $C_f$ are constants for $f_{flex}$.

A torsional ($f_{tors}$) and flexural ($f_{flex}$) resonance frequency are measured with the RUS and then applied against Eqs. 1-2. A unique combination of E and v does not represent the measured torsional or flexural frequency *singly* since there is actually a locus of E-v pairs that does. However, a unique E-v combination *does* result when those two resonances are combined, see Fig. 2. For example, the locus of E and v that satisfies a torsional frequency of 368 kHz intersects the locus of E and v that satisfies the flexural frequency of 465.4 kHz at one E and v pair - *it is that E and v pair that satisfy Eqs. 1-2 and that are the elastic properties of the test material and are the values that were determined for six different spherical WC materials.* That unique E and v pair, when re-inputted in the modal analysis, will predict resonant frequencies that match *all* the experimentally identified RUS peaks in a spectrum, such as shown in Fig. 1. This action serves as a verification of the determined E and v pair.

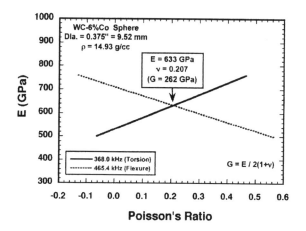

Figure 2. A unique combination of E and v (and G) satisfy both resonance frequencies.

In addition to the measurement of elastic properties, interest also existed in the present study to measure hardness of the WC compositions, relate it to the material's yield strength, and then use that value, the measured elastic properties, and FEA to estimate loads and velocities at which the WC spheres will undergo yielding and crack initiation. A material's yield strength can be estimated for many materials (metals particularly) by dividing its measured hardness by a factor of 2.6 to 3.0 [2] - a value of 3.0 is used in the present study. That yield strength can then be used as a material property in FEA to model the stress fields and to identify what compressive loads are necessary to initiate deformation in a sphere in contact with a target (whose material properties are also known). That information can then be used in conjunction with established elastic impact theory to determine what impact velocity is associated with the elastic limit of the material in the sphere according to |3|

$$P = \frac{4}{3}\sqrt{RE^*}\left(\delta_z\right)^{\frac{3}{2}} \quad , \tag{3}$$

where P is the compressive load, $1/R = 1/R_s + 1/R_t$, ($R_s$ is the sphere radius, $R_t$ is the target radius) and is equal to infinity for a flat surface, $E^*$ is the effective Young's modulus where $1/E^* = (1-v_s^2)/E_s + (1-v_t^2)/E_t$, $v_s$ and $v_t$ are Poisson's ratio of the sphere and target material, respectively, $E_s$ and $E_t$ are the Youngs moduli of the sphere and target material, respectively, and $d_z$ is displacement. For a sphere approaching a target at a velocity, V, the maximum compression during impact occurs at $d_z/dt = 0$, and it can be shown that V can be related to $d_z$ according to

$$\delta_z = \left(\frac{15mV^2}{16\sqrt{RE^*}}\right)^{\frac{2}{5}} \quad , \tag{4}$$

where $1/m$ is the effective mass, $1/m = 1/m_s + 1/m_t$, ($m_s$ is the mass of the sphere, and $m_t$ is the mass of the target and $1/m_t = 0$ for an "infinite" target mass). Equations 3-4 relate load to velocity when the sphere and target are linearly elastic. In parallel, FEA can be used to take into account yield behavior and identify the compressive load associated with the material elastic limit or where yielding initiates. Once identified, that load can be used in Eqs. 3-4 to determine the corresponding velocity.

To help with the understanding of the competition between yield initiation and the initiation of fracture processes (unlike ductile materials, brittle materials introduce the latter), the maximum 1st Principle tensile stress can be linked to ceramic tensile strength, and the loads and velocities to initiate cracking can be estimated and compared with loads and velocities that initiate yielding. Such an effort was undertaken in the present study.

II. EXPERIMENTAL PROCEDURES

Three WC sphere materials were received from the US Army Research Laboratory, and their designations and nominal sizes are listed in Table I. Diameters were measured to a resolution of 0.001 mm and densities were measured to a resolution of 0.01 g/cc for each.

Three-probe RUS analysis was performed on the spheres, and the spheres were subjected to excitation frequencies between 20-1000 kHz.

Table I.    Three WC composition and spherical diameter were examined.

| Composition | Nominal Diameter |
|---|---|
| WC - high purity (Roc500) | 0.250 in. (6.35 mm) |
| WC - 6%Co | 0.250 in. (6.35 mm) |
| WC - 12%Co | 0.250 in. (6.35 mm) |

Modal analysis was performed using ANSYS for all three sets. The measured ball diameter and density, and E and ν were inputted in the model where the latter two parameters were "bracketed" in the manner described in the Background section. Torsional and flexural frequencies were related to E and ν using multilinear regression according to Eqs. 1-2. An iterative analysis was performed where E and n were tandemly varied until a combination was found to satisfy both a torsional and flexural resonant frequency that was measured with the RUS. This iterative analysis determined E and ν (and G according to $G = E/2(1 + ν)$) in the manner illustrated in Fig. 2.

To work toward the estimation of the threshold loads and velocities that cause yielding and crack initiation in the WC spheres, the procedure illustrated in Fig. 3 was used. *Shock-induced effects were not considered in the analysis, and yield stress was assumed to be independent of loading rate in the analysis.* The average Vickers hardness at 500g was measured for each of the materials after their spheres were sectioned and metallographically polished. Yield stress was determined by dividing the measured average hardnesses by three. Finite element analysis was then performed using an axisymmetric model of a sphere coming into contact with a target and increasing the contact load between them. The measured elastic properties from the RUS experiments, the calculated yield stress, and the sphere geometry were combined to determine the radial, axial, hoop, shear (radial/axial plane), 1st Principle, and von Mises stress fields in the WC sphere as a function of compressive load.

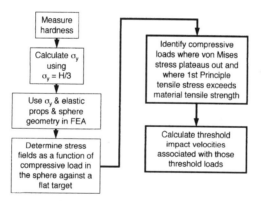

Figure 3.   Procedure used to calculate loads and velocities associated with yielding and crack initiation in the WC spheres.

The von Mises stress reaches a plateau value ($\approx \sigma_y$) when yielding commences; the load at which this plateau was attained was identified from FEA post-processing. Attention was paid to the magnitudes of the 1st Principle tensile stresses because (as a first approximation and neglecting strength-size-scaling issues and because the magnitude of tensile strength is typically an order of magnitude lower than compressive strength for the same ceramic material) crack or fracture initiation can be taken to commence when it exceeds the tensile strength of a monolithic ceramic. Consequently, the compressive load was identified that generated a $1^{st}$ Principle tensile stress that was greater than the tensile strength of the WC material. Lastly, with those two (yield and cracking) threshold loads identified, their corresponding velocity was calculated according to Eqs. 3-4.

## III. RESULTS & DISCUSSION

### Elastic Properties

The first (i.e., lowest frequency) torsional and flexural resonance frequencies were used to determine E and ν using the procedure previously described and illustrated in Fig. 2, and their results (to 3 significant digits) are summarized in Table II. Decreasing elastic modulus with increasing metal additive content was anticipated and observed. For all three WC sets, the identified E-ν pair was re-inputted in ANSYS with the corresponding sphere diameter and material density, and modal analysis re-performed to compare predicted resonance frequencies with those measured with the RUS. In every case, there was matching within 1% for all the peaks.

Table II. Resulting elastic constants from RUS analysis.

| Composition | Density (g/cc) | E (GPa) | ν | G (GPa) | K (GPa) |
|---|---|---|---|---|---|
| WC (Roc500) | 15.64 | 689 | 0.177 | 292 | 356 |
| WC - 6%Co | 14.93 | 635 | 0.196 | 266 | 348 |
| WC - 12%Co | 14.41 | 585 | 0.269 | 231 | 422 |

G = Shear Modulus = E/(2(1+ν)), K = Bulk Modulus = E/(3(1-2ν))

### Hardness and Yield Strength

Average hardnesses, H, for the three sets of WC are listed in Table III. Yield stress, $\sigma_y$, was calculated from each by dividing H by three, and these values are also listed in Table III. The trend in decreasing hardness with increasing metal additive content was expected. The Roc500 WC has less than 1% metal additive content and its hardness is almost double that of the WC composition with 12% cobalt additive. In turn, the calculated yield strength for the Roc500 WC is twice that of the WC-12%Co; however, as will be discussed below, the relatively low strength of the Roc500 likely results in an inability to ever attain or exploit its (potential) high yield strength. As will be observed too and unlike the Roc500, a high tensile strength of the WC-6% and WC-12% compositions coupled with lower yield strengths (compared to the Roc500) will likely result in their yielding *before* fracture processes initiate.

*Determination of Threshold Loads and Velocities*

An axisymmetric FEA model was used to model stresses in a WC ball. Elastic and yield properties of the Roc500, WC-6%Co, and WC-12% compositions with spherical geometry, and its contact against an infinitely stiff target and against a target with "hot-pressed-like SiC" properties were considered. The effect of friction can be accounted for in the model; however, the coefficient of friction was taken to be zero in this study. To assess the validity of the FEA model, an analysis was first performed by considering a linear elastic WC-6%Co sphere loaded in contact against a linear elastic hot pressed SiC target, and then comparing the resulting average axial compressive stresses as a function of compressive load against those calculated from the classical Hertzian solution. The model was concluded to be satisfactory as the differences between the modeled and Hertzian average contact stresses were less than 3%.

The von Mises and $1^{st}$ Principle (tensile) stresses received the most attention, though several additional stress fields (e.g., radial, axial, hoop, and radial/axial shear) were acquired for the sake of completeness. As an illustration, the von Mises and $1^{st}$ Principle stress fields in the WC are shown in Fig. 4. At a compressive load of 100N, the maximum von-Mises stress (Fig. 4 - upper left) has not yet attained $\sigma_y$ and linear elasticity still exists. Additionally, the maximum $1^{st}$ Principle tensile stress (located at the sphere surface just outside the contact area shared between the sphere and target) has not exceeded the tensile strength of this material (~2000 MPa), so crack initiation has not commenced either at 100N (Fig. 4 - upper right). The situation is different at 2000N though. The maximum von Mises stress has already attained $\sigma_y$ and the sphere is likely undergoing yielding (Fig. 4 - lower left) *and* the maximum $1^{st}$ Principle tensile stress (Fig. 4 - lower right) has exceeded the tensile strength too, so crack initiation has likely occurred. The $1^{st}$ Principal tensile stress is comprised primarily of radial tensile stress (see Figs. 4. - upper and lower right for location on surface and orientation); consequently, if cracking were to occur, then a ring-like crack would be expected.

Maximum von Mises and maximum $1^{st}$ Principle tensile stresses were determined as a function of load and their relationships are illustrated in Fig. 5. To compare material effects, this same diameter was used with loading against an infinitely stiff target. The compressive load where the von Mises stress was finally equivalent to the yield strength was easily identified (Fig. 5 - upper left). Additionally, the compressive load where the $1^{st}$ Principle tensile strength was finally equal to or exceeded the material tensile strength (estimated to be 600 MPa for the Roc500 WC and 2000 MPa for both the WC-6%Co and WC-12%Co materials) was easily identified as well (Fig 5 - upper right). These results (summarized in Table IV) show that the compressive loads for yielding in the WC-6%Co and WC-12%Co materials are lower than the compressive loads needed to initiate cracking in them (i.e., they will likely yield *before* fracture processes initiate in them) whereas the opposite is the case for the Roc500 composition; namely, cracking or fracture processes would occur *before* yielding.

Similarly, yielding in the WC-6%Co and WC-12%Co materials is likely to occur at lower velocities than fracture processes would (Figs. 5 - lower left and right). Conversely, fracture processes will initiate in Roc500 at lower velocities than yielding would. See Table IV.

The effect of target material properties on the yielding and crack initiation in a WC sphere was parametrically explored by compressively loading a 0.25 in. diameter WC-6%Co sphere against a hot-pressed SiC (i.e., E = 450 GPa, $\nu$ = 0.16, H = 20 GPa, $\sigma_y$ = H/3 = 6.67 GPa) target and comparing the response of the sphere against an infinitely stiff target as discussed above. As expected, the loads (Figs. 6 - upper left and right) and velocities (Figs. 6 - lower left and right) to initiate yielding and cracking in the WC sphere increase against a target with lower elastic modulus. Against the hot-pressed SiC target, the threshold loads increased by ~ 3.5-4x to initiate yielding and cracking, and the threshold velocities increased by ~ 4-5x compared with the same sphere contacting an infinitely stiff target. These results are tabulated in Table V.

These results show that yielding could occur in a 0.25 in. diameter WC-6%Co sphere impacting a hot-pressed SiC target at a velocity as low as ~ 1.5 m/s and that cracking could initiate at ~ 3.1 m/s. Though a similar exercise against a SiC target was not performed with a Roc500 or WC-12%Co sphere, their performances against an infinitely stiff target shown in Fig. 5 suggests that a 0.25 in. Roc500 WC sphere would experience crack initiation at a velocity less than 1.5 m/s, and that a WC-12%Co would experience yielding at a velocity slightly less than 1.5 m/s and crack initiation at a velocity slightly higher than 3.1 m/s.

As a last exercise, the threshold loads and velocities for a hot-pressed SiC target were examined concurrent with thresholds in an impacting WC-6%Co sphere, and a comparison of their values can be made in Table VI. Assuming a tensile strength of 800 MPa for the SiC, examination of the thresholds during impact suggests that: the hot-pressed SiC target will first start to experience cracking at 0.46 m/s; the WC sphere will then start to experience yielding at 1.51 m/s; the hot-pressed SiC will start to experience yielding at 2.12 m/s; and the sphere will start to experience cracking at 3.06 m/s.

## IV. SUMMARY

The Young's modulus of Roc500 WC, WC-6%Co, and WC-12%Co were experimentally determined to be 689, 635, and 585 GPa, respectively; with their Poisson's ratios of 0.177, 0.196, and 0.269, respectively; and average Vickers hardnesses are 24.0, 16.3, and 12.3 GPa, respectively. These are sufficient data to estimate threshold loads and velocities of yielding within a sphere made from these materials and in contact with a target material.

Yielding and crack initiation within a WC sphere that is impacting a hot-pressed SiC target are likely to occur at relatively low velocities (~ 1.5 and 3 m/s).

During impact, WC-6%Co and WC-12% spheres will likely experience yielding before fracture processes initiate within them. Conversely, Roc500 WC spheres will likely experience cracking first.

The 1st Principle tensile stress in a sphere in contact with a target is primarily a result of radial tensile stress - if cracking was to initiate on account of it, then a ring-like crack would likely be the pattern.

Roc500 WC would likely yield at much higher velocities than WC-6%Co and WC-12% balls; however, the low strength of Roc500 WC inhibits the ability to ever achieve that state or exploit that high yield strength.

## ACKNOWLEDGEMENTS

The author expresses appreciation for the: TARDEC-sponsorship under D. Templeton; assistance provided by ORNL's T. Kirkland and M. Ferber with the RUS measurements; for the hardness measurements generated by ORNL's S. Waters; and for helpful discussions with USARL's M. Normandia.

REFERENCES

[1]  A. A. Wereszczak, R. H. Kraft, and J. J. Swab, "Flexural and Torsional Resonances of Ceramic Tiles via Impulse Excitation of Vibration," *Ceramic Engineering and Science Proceedings*, **24** 207-13 (2003).
[2]  D. Tabor, *The Hardness of Metals*, Clarendon Press, Oxford, United Kingdom, 1951.
[3]  K. L. Johnson, *Contact Mechanics*, Cambridge University Press, Cambridge, United Kingdom, 1985.

Table III.  Average Vickers hardness (500g) and calculated yield strength.

| Composition | H (ave) (GPa) | $\sigma_y$ (GPa) |
|---|---|---|
| WC (Roc500) | $24.0 \pm 1.0$ | 8.00 |
| WC - 6%Co | $16.3 \pm 0.4$ | 5.43 |
| WC - 12%Co | $12.3 \pm 0.6$ | 4.10 |

5 indents comprise the average H, and shown ± values are standard deviations.
Calculated yield stress = $\sigma_y$ = H/3.

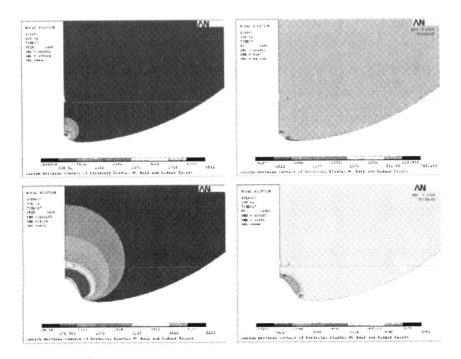

Figure 4.   Von Mises stresses were examined in regards to yield initiation within the spheres whereas the magnitude of the 1st Principal tensile stresses was examined for crack initiation. The maximum von Mises stress (upper left) and maximum 1st Principle tensile stress (upper right) at 100N were both low so no yielding or cracking was expected. At 2000N, both the von Mises (lower left) and 1st Principle tensile stresses (lower right) had sufficiently high magnitudes for yielding and crack initiation in the WC sphere. The target material is not shown to aid in the examination of the stresses in the sphere. The degree of contact-induced displacement is exaggerated in these images.

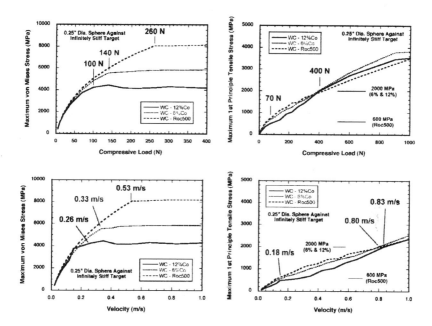

Figure 5. The maximum von Mises stress attained asymptotic values (i.e., yielding) at the shown compressive loads (upper left) and the maximum 1ˢᵗ Principle stresses reached tensile strengths at the shown compressive loads (upper right). Analogous threshold velocities for yielding (lower left) and crack initiation (lower right) are shown.

Table IV. Summary of predicted compressive loads and impact velocities that initiate yielding and crack initiation *in a 0.25 in. diameter sphere* (i.e., not the target) made from each of the three WC compositions.

| Composition | $P_{yield}$ (N) | $P_{crack}$ (N) | $V_{yield}$ (m/s) | $V_{crack}$ (m/s) |
|---|---|---|---|---|
| WC (Roc500) | 260 | 70 | 0.53 | 0.18 |
| WC - 6%Co | 140 | 400 | 0.33 | 0.80 |
| WC - 12%Co | 100 | 400 | 0.26 | 0.83 |

The following were assumed or used in the FEA to estimate these threshold values: an infinitely stiff target; static analysis; WC hardnesses and yield stresses listed in Table III; a tensile strength of 2000 MPa for both the 6% and 12% compositions and a tensile strength of 600 MPa for the Roc500 composition, and; H, $\sigma_y$, and $S_{ten}$ independent of loading rate.

Table V.  Comparison summary of predicted compressive loads and impact velocities that initiate yielding and crack initiation *in a 0.25 in. diameter sphere* (i.e., not the target) made from the WC-6%Co composition (E=635 GPa, $\nu$=0.196, H=16.3 GPa, and $\sigma_y$=H/3=5.43 GPa) against an infinitely stiff target and against a hot-pressed SiC target having the following properties: E=450 GPa, $\nu$=0.16, H=20 GPa, and $\sigma_y$=H/3=6.67 GPa.

| ------ Infinitely Stiff Target -------- | | | | -------- Target (hot-pressed SiC) --------- | | | |
| $P_{yield}$ (N) | $P_{crack}$ (N) | $V_{yield}$ (m/s) | $V_{crack}$ (m/s) | $P_{yield}$ (N) | $P_{crack}$ (N) | $V_{yield}$ (m/s) | $V_{crack}$ (m/s) |
|---|---|---|---|---|---|---|---|
| 140 | 400 | 0.33 | 0.80 | 600 | 1400 | 1.51 | 3.06 |

A loading-rate-independent tensile strength of 2000 MPa was used in the P and V estimations of cracking.

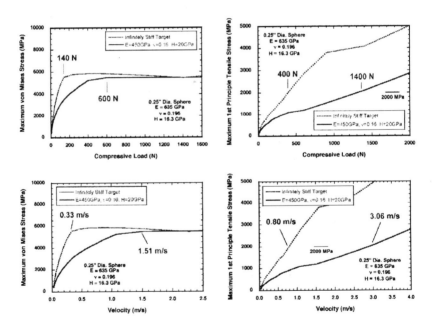

Figure 6.  Upon comparison of impact against an infinitely stiff target and a target having SiC properties, both the loads and velocities to initiate yielding and cracking in a WC sphere increase.

Table VI. Comparison summary of predicted compressive loads and impact velocities that initiate yielding and crack initiation in a 0.25 in. diameter sphere made from the WC-6%Co composition (E=635 GPa, $\nu$=0.196, H=16.3 GPa, and $\sigma_y$=H/3=5.43 GPa) and in a hot-pressed SiC target having the following properties: E=450 GPa, $\nu$=0.16, H=20 GPa, and $\sigma_y$=H/3=6.67 GPa.

| --------- Sphere --------- | | | | --------- SiC Target --------- | | | |
|---|---|---|---|---|---|---|---|
| $P_{yield}$ (N) | $P_{crack}$ (N) | $V_{yield}$ (m/s) | $V_{crack}$ (m/s) | $P_{yield}$ (N) | $P_{crack}$ (N) | $V_{yield}$ (m/s) | $V_{crack}$ (m/s) |
| 600 | 1400 | 1.51 | 3.06 | 900 | 150 | 2.12 | 0.46 |

A loading-rate-independent tensile strength of 2000 MPa and 800 MPa was used in the P and V estimations of cracking in the WC and SiC, respectively.

# ON THE ROLE OF IMPACT DAMAGE IN ARMOR CERAMIC PERFORMANCE

Joseph M. Wells
JMW Associates, 102 Pine Hill Blvd, Mashpee, MA 02649-2869
(508) 477-5764, E-mail jmwconsultant@comcast.net

## ABSTRACT

The scientific and engineering prediction of the effective ballistic performance of armor targets is highly desirable to expedite both the development of improved armor ceramics and their incorporation into advanced armor systems. The prediction and assessment of ballistic performance is most often focused on the macro-penetration phenomenon, which is relatively straight forward to observe and measure. Unfortunately, the cognition of the discrete intrinsic material and/or extrinsic target architectural design factors necessary for mitigating and/or controlling the penetration process in notional light-weight passive armor has proven quite elusive. Certainly, it is well known that significant physical damage results from ballistic impact in addition to any projectile penetration. In fact, in armor ceramics, complex internal damage, at both the micro- and the meso/macro-scale, is observed even in the complete *absence* of any penetration. Furthermore, since such appreciable impact damage occurs *prior* to the onset of penetration in armor ceramics, the question then needs to be addressed as to what role(s) that existing impact damage would have on the initiation and progression of subsequent penetration. Moreover since the ceramic damage and the penetration processes are both dissipative of kinetic energy transferred from the impacting projectile, the overall ballistic performance of impacted ceramics must be related to both impact damage and penetration. Various issues are raised and discussed relating to the characterization and utilization of actual impact damage in the evaluation of the overall performance of armor ceramics.

## INTRODUCTION

Armor ceramics are a critical engineering material in legacy, current and future vehicle and personnel armor concepts and designs. After several decades of incremental advancements in armor performance resulting from extensive empirical experimentation with both ceramic materials and architectural designs, one is still lacking rigorous technical models to accurately predict and guide the direction(s) necessary to achieve the affordable, light weight, and high performance ceramic armor required by modern warfare[1-5]. Thus, the perception arises that the technical paradigm for accelerating armor ceramic development and implementation very much needs to be substantially improved.

Historically the penetration phenomenon has been the predominant focus of most of the scientific and engineering studies in the terminal ballistics arena. Unfortunately, considerably less emphasis has been focused on the physical damage occurring during and resulting from ballistic impact. Among the several reasons for this situation has been the appreciably greater difficulty in the dimensional and morphological characterization of impact damage versus penetration. Additionally, since impact damage is essentially an asymmetric 3-D phenomenon, whereas the penetration phenomenon has frequently been quite adequately represented in 2-D, the desired 3-D predictive modeling of impact volumetric damage is substantially more challenging. Nevertheless, the desirability and feasibility of predictive impact damage modeling is becoming increasing more apparent. Consequently, the question is raised as to what role existing impact damage would have on the initiation and progression of subsequent penetration?

To begin to answer this important question, one needs to expand our collective knowledge and understanding of the actual damage occurring in impacted ceramic targets before we proceed to the understanding of their influence on subsequent and concurrent penetration. First, it is helpful to state that what is meant here by impact damage are the irreversible impact-induced physical modifications of the material shape, continuity, and structural integrity. Relevant manifestations of such physical damage features in armor ceramics include cracking of various forms (ring, radial, conical, laminar, etc.) on various size scales, as well as surface craters and surface debris flow features, the near sub-surface comminuted zone, impact induced porosity or voids, penetration cavity formation, inhomogeneous deformation and fragmentation, and imbedded residual projectile fragments. Second, it is essential to employ volumetric NDE diagnostic capabilities for the detection, characterization, visualization and analysis of such impact damage features to identify, understand and eventually model these complex impact damage features.

The author and his collaborators[6-14] have utilized the unique noninvasive 3-D capabilities of x-ray computed tomography, XCT, toward this end and have made significant progress on improving our understanding of actual volumetric ceramic impact damage. Certainly much more remains to be done in further developing this damage diagnostic area combined with developing knowledge and understanding related to the specific ceramic material / ballistic conditions under which such damage occurs. Another broader key concern is what technical strategy and methodology will incorporate this volumetric damage knowledge to elucidate how ballistic impact damage actually influences penetration and the overall ballistic performance? It is not apparent that such a suitably coordinated strategy and methodology exist at the present time. Ultimately, it is desired to develop predictive mechanics-based material damage models that would allow the pre-impact analytical screening of notional and developmental armor ceramic material/architectural modifications prior to their expensive fabrication, ballistic testing, and verification. While it is premature to attempt to provide the answers to these difficult questions here, the intended contribution of this paper is rather to introduce and discuss some of these relevant questions and issues in the hope of stimulating a more thorough subsequent exploration and evaluation.

BACKGROUND

While resistance to penetration is essential in an effective passive armor material system, still not well understood at present, conceptually or practically, is how to modify engineering ceramic materials to significantly improve their penetration resistance. Rule of thumb experience has indicated that high hardness, high compressive strength, high toughness, low density and low cost are very desirable attributes for practical armor ceramics. Yet such combined ensemble attributes are difficult to attain, and collectively still provide no assurance of being sufficient for maximum desired impact protection. Historically, considerable attention has been placed on ballistic experiments where the rate and extent of penetration has been measured, modeled and computer simulated. Such experiments provide an empirical assessment of the penetration resistance capability of a specific material/target configuration, but generally do not provide substantial understanding of the processes by which the armor ceramic material itself might be modified to effectively resist that penetration.

Upon impact, and prior to and during penetration, armor ceramic materials experience considerable physical damage including comminution (pulverizing) of the ceramic volume directly beneath the impinging projectile, and extensive micro- and meso-cracking beyond this comminuted zone. Since the resulting individual damage features reflect both the intrinsic material damage processes and the impact induced stress-strain-time conditions, they may well provide suggestions for the eventual means of "damage control/damage mitigation". For example, if one can suppress, mitigate, localize or otherwise control the evolution of critical damage morphologies (by either extrinsic target modifications and/or intrinsic ceramic material modifications), it is likely that penetration may be impeded and perhaps even prevented. One well known example of limiting impact damage by *extrinsic* means, and consequently not only limiting but actually preventing penetration, is the application of substantial physical constraint to produce "interface defeat" as demonstrated by Hauver et al[15]. Decreased impact damage and penetration with much less surrounding constraint material was subsequently demonstrated by Bruchey and Horwath[16] using a welded Ti-6Al-4V encapsulation package and then dramatically further by Rupert et al[12] by applying a compressive prestress via shrink fitting a modest 17-4 PH steel ring around a cylindrical $TiB_2$ ceramic target prior to encapsulation. Subsequent results on modeling the effects of extrinsic prestress by Holmquist and Johnson[17] also reported that prestressing enhanced the ballistic performance of ceramics. Recently, Holmquist et al[18] reported on the achievement of significantly extended dwell conditions (indicating reduced damage evolution) employing only a surface copper buffer plate and without the use of any extrinsic target confinement material.

Similarly effective *intrinsic* approaches will most likely depend on a significantly improved awareness and knowledge of actual impact damage details and the creative material design and processing approaches by which such damage features can be suppressed, mitigated, localized or otherwise controlled. Obviously, one prefers to start with the best quality (minimum presence of pre-existing as-manufactured defects) ceramic materials commercially practical. While this may be necessary, it is not sufficient as impact damage will still occur. New innovative engineered ceramic materials, or combinations thereof, unquestionably need to be designed and economically produced which are significantly more damage tolerant. This, of course, is more easily stated than accomplished. Predictive mechanics-based damage models would be most helpful in the development and pre-test screening of the various candidate material schemes proposed for increased ballistic performance. For much of the above to occur, we find ourselves returning back to many basic questions that come to mind concerning *the role of impact damage in armor ceramic ballistic performance*. A few representative examples of such relevant questions are:

- What volumetric damage features are created by ballistic impact?
- How can impact damage be diagnosed, visualized & quantified in 3-D?
- Can real damage be adequately characterized for use in damage-based predictive modeling?
- How can predictive models effectively include such damage characterizations?
- How does impact damage affect the subsequent penetration process?
- How can impact damage be mitigated and/or controlled to significantly improve ballistic performance?
- What needs to be done to answer these questions?
- What are the necessary goals and actions – short – medium – long term?
- Who will lead the way? Who will collaborate? Who pays?
- Where do we go from here?

Answers to all of these relevant questions are far beyond the scope and intent of the present paper. The first three are raised and discussed here selectively and briefly in the allowable space to stimulate further collective creative thinking and planning towards the progressive (and ultimate) resolution of those questions remaining.

## WHAT VOLUMETRIC DAMAGE FEATURES ARE CREATED BY BALLISTIC IMPACT?

Much of the limited effort to date to examine the damage occurring in the impacted ceramic has been hampered either by the shattering and dispersion of unconstrained ceramic targets or by the practical limitations associated with destructive sectioning of impacted encapsulated ceramics. Destructive sectioning is costly and time consuming, and introduces the possibility of irreversible extrinsic damage during the destructive sectioning process itself. Another liability of destructive sectioning is the limitation of achieving only an observed 2D planar perspective of the damage morphology on the plane of sectioning with little, if any, additional information available regarding the unobserved contiguous damage which is obscured by the opacity of the adjacent bulk ceramic. This latter limitation, can easily lead to unwarranted assumptions regarding the volumetric symmetry of the total damage morphology. Nevertheless, traditional ceramographic destructive sectioning has provided the original baseline of specific damage features such as radial, ring or circular, lateral or laminar and Hertzian conical cracking often observed and reported for impacted ceramics. Figure 1 shows an excellent recent example of these cracking types demonstrated by LaSalvia et al [19]. It should be noted that, while quite graphic, such cracking damage indications are limited to a 2D perspective only. Interesting high speed Edge-on-Impact (EOI) damage observations by Strassburger et al [20-22] in transparent target materials are quite revealing but experience superposition of both surface and interior damage indications and are limited to 2D planar observations as well.

Fountzoulas et al[23] recently reported on the 3D numerical simulation and prediction of impact damage patterns in SiC$_N$ ceramic using such 2D damage observations with the

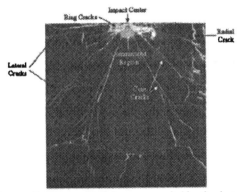

Figure 1. Traditional forms of impact cracking damage revealed by the destructive examination in a SiC$_N$ armor ceramic by J.C. LaSalvia et al[19].

phenomenological Johnson-Holmquist model. Several additional manifestations of armor ceramic damage including surface ring steps, near-surface radial expansion and shallow penetration radial cracking, spiral and concentric "hourglass-shaped" ring cracking morphologies, impact induced porosity localization about cracking indications, and residual penetrator fragment distributions have been revealed in 3D perspective by Wells et al.[6-14] using virtual nondestructive industrial x-ray computed tomography, XCT, diagnostic techniques. Figure 2 provides a comparative overview of these several impact damage features revealed by these different approaches.

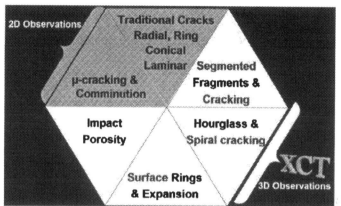

Figure 2. Comparative Overview of various 3D impact damage features detected and visualized with XCT versus 2D ceramographic observations.

## HOW CAN IMPACT DAMAGE BE DIAGNOSED, VISUALIZED & QUANTIFIED IN 3-D?

Conventional NDE modalities, such as ultrasound, eddy current, thermal wave imaging, and even 2D high speed flash x-rays, have not yet demonstrated an acceptable level of impact damage characterization and 3D visualization. Consequently, it is considered beneficial to utilize

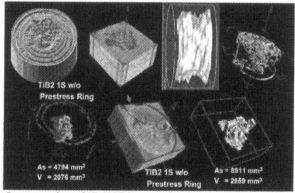

Figure 3. Summary view of 3D XCT images of various impact damage features.

the noninvasive and direct post-impact XCT imaging diagnostic approach to reveal, characterize, and quantify the true 3D morphological details of actual impact damage throughout the entire target ceramic volume. While there is inadequate space here to review in detail the diagnostic capabilities of XCT for this purpose, many relevant examples have been previously demonstrated by the author and his collaborators [6-14]. A few summary graphics highlighting some of these examples capabilities are shown in figures 3 and 4.

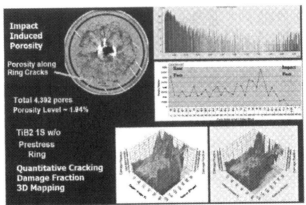

Figure 4. Summary of 3D XCT Impact Damage Quantification Examples.

## CAN REAL DAMAGE BE ADEQUATELY CHARACTERIZED FOR USE IN DAMAGE BASED PREDICTIVE MODELING?

Certainly, impact damage diagnostic capabilities have improved dramatically with XCT in the past decade or less. At a minimum, the existence and the complexities of several impact damage features beyond those typically described with 2D ceramographic techniques have been demonstrated as shown in figures 5, 6, & 7. Furthermore, the feasibility of obtaining quantitative

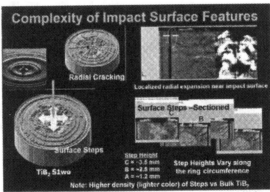

Figure 5. Summary image of demonstrated XCT diagnostic capabilities related to impact surface damage features of interest.

details of many of these 3D damage features has also been established with non-intrusive metrology. Whether or not these diagnostic capabilities are yet useful and sufficient at this stage for damage-based predictive modeling remains an open question since such modeling capability remains quite embryonic at present. Where then do we go from here?

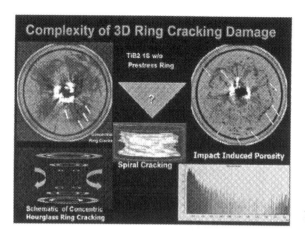

Figure 6. Summary image of demonstrated XCT diagnostic capabilities related to impact ring cracking damage complexities.

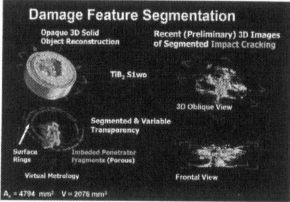

Figure 7. Summary image of demonstrated XCT diagnostic capabilities related to the segmentation (isolation) of various impact damage features.

One suggestion involves the aggressive incorporation of the relatively recent, and well documented, innovative XCT capabilities mentioned above to characterize, visualize and analyze actual (not simulated) ceramic impact damage. Realistically, further refinements in the development and application of the XCT damage assessment technology are anticipated and desired. Hopefully, other useful diagnostic techniques may also become available as well. Moreover, for impact damage to be systemically included in the assessment of armor ceramic performance, a greatly improved interdisciplinary synergy of several related technical areas is required. Such areas include damage feature analyses and correlations from laboratory screening tests for mechanical properties, quasi-static instrumented indentation testing, sub-scale variable velocity laboratory impact tests, and so forth. Furthermore, considerable attention also needs to be focused on the strategy and methodology for effectively using such evolving and previously unavailable damage characterization information and knowledge. With the resultant improvements in the characterization and understanding of the effects of such damage on penetration and overall ceramic armor performance, one would have an improved pragmatic rationale with which to address ceramic material modifications and architectural design improvements.

## FACTORS INHIBITING DAMAGE-BASED CERAMIC PERFORMANCE MODELS

Since the concept of damage-based ceramic modeling is not new [1,2], one has to reflect on why cumulative progress achieved to date has been so limited and not particularly realistic and useful for developing improved armor ceramics? Unquestionably, the development of such viable damage-based performance modeling is a difficult problem. Nevertheless, it does not necessarily appear to be an insolvable one. Several relevant factors which likely have been contributing to the inhibition of progress in this direction include:

- Lack of an apparent broad-based recognition and acknowledgement that impact damage may be equally as critically important as penetration for performance assessment.
- Real damage is a 3-dimensional phenomenon and is often inadequately described by commonly utilized 2D diagnostic approaches.
- Actual impact damage is frequently very complex, inhomogeneous, and asymmetrical; it is neither easily characterized physically nor described mathematically.
- Impact damage features occur on both micro- and meso-size scales.
- Various damage features often occur in a superimposed "tangle" of damage forms which are not easily recognized nor separated.
- Extensive destructive examination is expensive, incomplete and physically unforgiving.
- Traditional NDE modalities have not, to date, demonstrated accurate and high resolution volumetric characterizations of actual armor ceramic impact damage.
- Other significant ceramic damage features, beyond traditional cracking modes, have been observed and may need to be included in future damage modeling efforts. Such features include impact induced porosity, concentric "hourglass-shaped" ring crack morphologies, spiral cracking morphologies, surface deformations and intermixing of penetrator debris with surface ceramic particles, residual penetrator fragments, etc.
- Promising 3D XCT damage diagnostic and characterization capabilities and results to date are not widely recognized and utilized.

In general, for some time there does not appear to have been a concerted effort to revisit the impact damage-based modeling potential to advance the armor ceramic performance issues. Hopefully, that may change in the near future. Some suggestions to assist in the furtherance of this potential are briefly discussed in the following section.

## SUGGESTIONS FOR FURTHER ACTION

Among the many issues which need to be addressed to advance the development of significant engineering improvements in armor ceramic performance, one needs to explore the premise that impact damage is an active material related factor significantly influencing the penetration phenomenon. As logical and direct as this premise appears, one cannot but help observe the overwhelming preponderance of traditional and current research emphasis solely on penetration without the explicit and interactive inclusion of impact damage considerations. What this suggests is that a concerted effort needs to be championed to motivate and organize a renewed collaborative effort to address the damage premise.

Suggested objectives for such a renewed collaborative damage effort might well include:

- Increasing community acceptance of the need to vigorously explore the role of impact damage details in the overall performance of armor ceramics.

- Incorporation of 3D NDE diagnostic capabilities for high resolution damage characterization, visualization, and assessment into predictive ballistic impact damage modeling [24-24].

- Establish meaningful correlations between the 3D damage characteristics created under quasi-static mechanical and instrumented indentation screening tests and high strain rate impact screening tests compared to those developed in various ballistic impact tests.

- Determine the volumetric details of specific damage features evolving as a function of increasing impact severity up to and beyond the penetration threshold.

- Develop mathematical representations of actual damage features compatible with requirements for a damage-based predictive modeling methodology.

- Define and explore intrinsic ceramic material modifications so as to suppress, mitigate, localize or diffuse impact damage accumulations to below the severity levels associated with catastrophic loss of the structural integrity of the bulk ceramic.

- Develop a mathematical model for the maximum dynamic stress state sustainable in the presence of a specific damage condition before ceramic structural instability occurs and penetration commences.

- Couple evolving ceramic bulk damage characterization knowledge with the interest on resistance to comminution and resistance to penetration through the comminuted ceramic particles.

- Collaborative exploration of improved impact damage-based modeling by incorporating evolving 3D damage characterization knowledge with advanced computational modeling software and hardware capabilities.

## SUGGESTED MILESTONES FOR FURTHER PROGRESS

A focused technical exploration and collaborative interaction of impact damage characterization and assessment might be approached along the lines indicated by the schematic outline shown in figure 8. Suggested milestones indicating acceptable progress might be indicated by the following:

- Understanding the nature, morphology, extent and consequences of the 3D ceramic damage features occurring under various architectural target design concepts and ballistic impact conditions.

- Relating dwell transitional effects to 3D impact damage extent and morphology leading up to the initiation and progression of penetration.

- Defined contributions of both micro- and meso-scale damage elements on the structural integrity of the target ceramic to resist and/or prevent penetration.

- Develop a structural damage containment rationale for future ceramic material/design architectural modifications to mitigate the penetration process.

- Predictive damage modeling capable of evaluating notional and developmental armor ceramic modifications.
- Recalibration of conventional NDE modalities to accommodate rapid and field portable inspection techniques for the assessment of impact damage in armor ceramics.

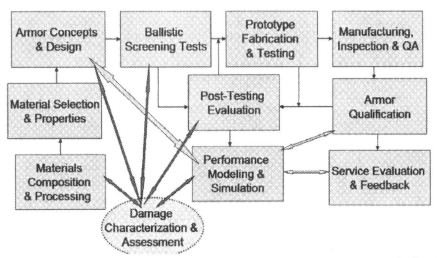

Figure 8. Schematic illustrating suggested application areas for impact damage characterization & assessment in the evolution of new armor materials.

SUMMARY

The time is right to revisit the role of impact damage on the ballistic performance of armor ceramics. Several questions are posed above to stimulate and encourage the collective reconsideration and discussion of how impact damage specifically influences penetration, and consequently the overall performance of armor ceramics.

The premise is advanced here that increasingly severe impact damage progressively reduces the structural integrity of the ceramic target and consequently affects the onset and extent of subsequent penetration. If substantiated, and if appropriate means can be devised to effectively model the prevention, mitigation, and/or morphological control of such damage within the ceramic, the significant potential exists for reducing penetration and thus, improving the overall ballistic performance. Eventually, with better impact damage characterization, visualization, and analysis, and with improved modeling capabilities of the detailed effects of impact damage on ceramic armor penetration, an improved technical rationale becomes available with which to systematically address notional and developmental armor ceramic materials and architectural design modifications. Such a paradigm shifting rationale appears timely and necessary to achieve significant improvements in our understanding and in the practical achievement of the overall desired performance of armor ceramics. It remains to be seen how the ballistic community will respond to the above.

ACKNOWLEDGEMENTS

Grateful acknowledgements are extended to Drs. David Stepp, Doug Templeton, Leo Christodoulou, Bill Bruchey, Jerry LaSalvia, Steve Cimpoeru and Mr. Nevin Rupert for their advice and encouragement in the furtherance of the pursuit of the role of impact damage on the performance of armor ceramics.

REFERENCES

1. D.M. Stepp, "Damage Mitigation in Ceramics: Historical Developments and Future Directions in Army Research," *Ceramic Armor Materials by Design*, Ed. J.W. McCauley et. al., Ceramic Transactions, v. 134, ACERS, 421-428 (2002).
2. A.M. Rajendran, "Historical Perspective on Ceramic Materials Damage Models," **ibid**, 281-297 (2002).
3. D.W. Templeton, T.J. Holmquist, H.W. Meyer, Jr., D.J. Grove and B. Leavy, "A Comparison of Ceramic Material Models," **ibid**, 299-308 (2002).
4. J.C. LaSalvia, "Recent Progress on the Influence of Microstructure and Mechanical Properties on Ballistic Performance," *ibid*, 557-570 (2002).
5. J.M. Winter, W.J. Bruchey, J.M. Wells, and N.L. Rupert, N.L. "Review of Available Models for 3-D Impact Damage," Proceedings 21st Intn'l Symposium on Ballistics, Adelaide, ADPA, v.1, 111-117 (2004).
6. J. M. Wells, "Progress on the NDE Characterization of Impact Damage in Armor Materials," Proceedings of 22nd IBS, ADPA, v2, 793-800 (2005).
7. J.M. Wells, "On Non-Destructive Evaluation Techniques for Ballistic Impact Damage in Armor Ceramics," Ceramic Engineering and Science Proceedings, v26, [7], 239 – 248, (2005).
8. H.T. Miller, W.H. Green, N. L. Rupert, and J.M. Wells, "Quantitative Evaluation of Damage and Residual Penetrator Material in Impacted TiB2 Targets Using X-Ray Computed Tomography," 21st International Symposium on Ballistics, Adelaide, Au, ADPA, v1, 153-159 (2004).
9. J. M. Wells, N. L. Rupert, and W. H. Green, "Progress in the 3-D Visualization of Interior Ballistic Damage in Armor Ceramics," Ceramic Armor Materials by Design, Ed. J.W.McCauley et. al., Ceramic Transactions, v134, ACERS, 441-448 (2002).
10. J. M. Wells, N. L. Rupert, and W. H. Green, "Visualization of Ballistic Damage in Encapsulated Ceramic Armor Targets," 20th Intn'l Symposium on Ballistics, Orlando, FL, ADPA, v1, 729-738 (2004).
11. W. H Green, K.J. Doherty, N. L. Rupert, and J.M. Wells, "Damage Assessment in TiB2 Ceramic Armor Targets; Part I-X-ray CT and SEM Analyses," Proc. MSMS2001, Wollongong, NSW, AU, 130-136 (2001).
12. N.L. Rupert, W.H. Green, K.J. Doherty, and J.M. Wells, "Damage Assessment in TiB2 Ceramic Armor Targets; Part II - Radial Cracking," Proc. MSMS2001, Wollongong, NSW, AU, 137-143 (2001).
13. J.M. Wells, W.H. Green, and N.L. Rupert, "Nondestructive 3-D Visualization of Ballistic Impact Damage in a TiC Ceramic Target Material," Proceedings MSMS2001, Univ. of Wollongong, NSW, AU,159-165 (2001).
14. J.M. Wells, W.H. Green, and N.L. Rupert, "Nondestructive 2D and 3D Visualization of Interface Defeat Based Ballistic Impact Damage in a TiC Ceramic Target Disk," Proceedings 22nd Army Science Conference, Baltimore, MD, 538-545 (2000).
15. G.E. Hauver, P.H. Netherwood, R.F. Benck, and L.J. Kecskes, "Ballistic Performance of Ceramic Targets," *Army Symposium on Solid Mechanics, USA*, (1993).
16. W.J. Bruchey and E.J. Horwath, "System Considerations Concerning The Development of High Efficiency Ceramic Armors," http://www.dtic.mil/matris/sbir/sbir012/a01-039a.pdf
17. T.J. Holmquist and G.R. Johnson, "Modeling Prestressed Ceramic and its Effect on Ballistic Performance," Intn'l J. of Impact Engineering, 31, 113-127 (2005).
18. T.J. Holmquist, C.E. Anderson, Jr., and T. Behner, "Design, Analysis and Testing of an Unconfined Ceramic target to Induce Dwell," Proceedings of 22nd IBS, ADPA, v2, 860-868 (2005).
19. J.C. LaSalvia, M.J. Normandea, H.T. Miller, and D.E. McKinzie, "Sphere Impact Induced Damage in Ceramics: I. Armor-Grade SiC and TiB2," Ceramic Engineering & Science Proceedings, v26, [7], 183-192 (2005).
20. E. Strassburger, "Visualization of Impact Damage in Ceramics Using the Edge on Impact Technique," *Int. J. Appl. Ceram. Technol., 1, [3], 235-42 (2004).*
21. E. Strassburger, P.Patel, J. W. McCauley, and D.W. Templeton, "High-Speed Photographic Study of Wave and Fracture Propagation in Fused Silica," Proceedings of 22nd IBS, ADPA, v2, 761-768 (2005).

22. E. Strassburger, P.Patel, J. W. McCauley, and D.W. Templeton, "Visualization of Wave Propagation and Impact Damage in a Polycrystalline Transparent Ceramic-AlON," Proceedings of 22nd IBS, ADPA, v2, 769-776 (2005).
23. C.G. Fountzoulas, M.J. Normandia, J.C. LaSalvia, and B.A. Cheeseman, "Numerical Simulations of Silicon Carbide Tiles Impacted by Tungsten Carbide Spheres," Proceedings of 22nd IBS, ADPA, v2, 693-701 (2005).
24. J.M. Wells, "Considerations on Incorporating XCT into Predictive Modeling of Impact Damage in Armor Ceramics," Ceramic Engineering and Science Proceedings, v26, Issue 7, 51-58, (2005).
25. J. M. Wells, "On Incorporating XCT into Predictive Ballistic Impact Damage Modeling," Proceedings of 22nd IBS, ADPA, v2, 1223-1230 (2005).

THE INDENTATION SIZE EFFECT (ISE) FOR KNOOP HARDNESS IN FIVE CERAMIC
MATERIALS

Trevor Wilantewicz and W. Roger Cannon
Rutgers, The State University of New Jersey
607 Taylor Road
Piscataway, NJ 08854

George Quinn
National Institute of Standards and Technology
Gaithersburg, MD 20899

ABSTRACT
    The variation of indentation hardness with contact load i.e., the indentation size effect (ISE),
was investigated for five ceramic materials: (i) AlON, (ii) AD995 CAP3 $Al_2O_3$, (iii)
pressureless-sintered SiC (Hexoloy SA), (iv) SiC-N, and (v) SiC-B. A Knoop diamond indenter
was used to make indentations in the load range from 0.49 N (0.05 Kg) to 137.3 N (14 Kg). The
Knoop hardness decreased approximately 26%, 38%, 36%, 32%, and 31% for the five materials,
respectively, over the entire load range. The Knoop hardness of all five materials continued to
decrease as the load was increased beyond 2 Kg. In addition, severe cracking around indentation
sites generally correlated with a lower Knoop hardness compared to less-cracked indentation
sites for the silicon carbide and AD995 CAP3 materials. Accurate hardness measurements are
necessary in order to detect variability in hardness not attributable to operator or instrument
differences, and which may yield insight into ballistic performance.

INTRODUCTION
    Previous work has shown that conventional indentation hardness values vary with the
magnitude of the applied load for both Knoop and Vickers indenter geometries.[1,2,3,4,5,6,7,8,9,10,11]
In general, the hardness decreases with increasing applied load. This phenomenon is called the
Indentation Size Effect (ISE). The effect is present in ceramics[1,2,3,4,5,6], metals[7,8], and
glasses.[9,10,11] Differential elastic recovery, cracking, frictional effects, surface residual stresses,
and operator mistakes are some factors which have been proposed to contribute to the effect.[1,7,9]
The ISE may manifest different damage and deformation processes during indentation
penetration and may provide valuable insights about the ballistic impact resistance of ceramic
armors.
    Quinn and Quinn[1] have shown the Vickers hardness of ceramic materials, including
aluminum oxide, silicon carbide, and silicon nitride, decreases as indentation load increases, and
then to level off to a more or less constant value at higher loads. The point at which the
transition to constant hardness took place was found to be linearly related to a brittleness index
which they defined as $HV_cE/K_{IC}^2$, where $HV_c$ is the Vickers hardness at the transition point i.e.,
the critical hardness, E is Young's modulus, and $K_{IC}$ fracture toughness.[1] Materials with lower
values of the brittleness index had transition points corresponding to larger indentation diagonal
sizes, as well as larger loads.[1]
    Swab[2] measured Vickers and Knoop hardness as a function of indentation load for seven
ballistic-grade ceramic materials including hot-pressed SiC, conventionally sintered $Al_2O_3$, two
types of hot-pressed $B_4C$, and three types of WC. Very little change in hardness with load was

found for the WC materials indented with the Vickers diamond. For the silicon carbide, aluminum oxide, and one of the boron carbide materials, hardness data could not be obtained at several different loads to generate a hardness-load curve due to extensive cracking around Vickers indentations interfering with diagonal length measurements. One might question whether cracked indentations are valid or not. Our stance is that some cracking is integral to the indentation process. Cracking is an important mechanism contributing to the deformation response of the material. Of course, if the cracking is so severe that it is impossible to see the indentation corners or the indentation shape, then the cracking should be considered excessive and the indentations deemed invalid. Guidance on this is included in ASTM standards C 1326[12] and C 1327[13] and ISO standard 14705[14] for ceramics.

Swab[2] was able to measure the Knoop hardness for all materials over the entire load range, due to significantly less cracking and the associated interference with the diagonal tips. The boron carbide materials were found to undergo the greatest hardness decrease ($\approx$45%) over the load range 0.98 N to 98 N, followed by the SiC and Al$_2$O$_3$ materials ($\approx$25% decrease), and lastly the WC materials ($\approx$13 to 20% decrease). In addition, Swab[2] found the Knoop hardness to decrease only slightly for loads greater than 1 Kg, and suggested a 2 Kg load as the minimum to use for Knoop hardness testing of hard armor ceramic materials. The current works objectives were to further investigate the load dependence of the Knoop hardness of five ceramic materials.

## EXPERIMENTAL PROCEDURES
### Materials

Manufacturer supplied samples of aluminum oxynitride (AlON), pressureless-sintered aluminum oxide (AD995 CAP3), hot-pressed (HP) alpha silicon carbide (SiC-N and SiC-B), and pressureless-sintered (PS) alpha silicon carbide ($\alpha$-SiC Hexoloy SA) in the form of 'b-size' bend bars were used in the investigation.[a] Table I lists some material properties supplied by the manufacturers of the material. The property data for the AlON was from Swab et al.[15]

### Sample Preparation

For the hardness testing, a small piece of material from each bar was cut using a diamond saw and subsequently mounted in epoxy for grinding and polishing. Samples were ground using a 20 $\mu$m diamond wheel, and subsequently polished with 15 $\mu$m, 9 $\mu$m, 6 $\mu$m, 3 $\mu$m, 1 $\mu$m, and 0.5 diamond solutions on a polishing wheel, yielding relatively smooth surfaces. The side opposite the specimen was ground flat using SiC paper to ensure a level specimen. The mounted specimens were then ultrasonicated in isopropyl alcohol for about ten minutes to remove any polishing residue.

### Hardness Testing

Indentations were made using a conventional hardness tester[b] equipped with a Knoop diamond pyramid. A specimen leveling device was used to hold the specimens perpendicular to the diamond indenter in order to ensure symmetric indentations and prevent sample movement during testing. A minimum of ten readable indentations were used to calculate the Knoop hardness at any particular load. The diamond indenter was new and care was used to make sure

---

[a] Certain commercial materials or equipment are identified in this paper to specify adequately the experimental procedure. Such identification does not imply endorsement by NIST and Rutgers University nor does it imply that these materials or equipment are necessarily the best for the purpose.
[b] Zwick Model 3212, Zwick USA, GA

no damage formed on the diamond during the course of testing. In addition, NIST standard reference material SRM 2830[16] was used to check the accuracy of the hardness tester, as well as that of the digital filar measuring system. The latter was accomplished by measuring the NIST-produced Knoop indentations on the SRM material, and comparing these values to the NIST-measured values. Indentations were made at the required 19.6 N (2 Kg) load in the SRM. All diagonal length measurements were made using the same digital filar measuring system, however at different magnifications in order to accommodate the different sized indentations of the hardness tests. Magnifications of 200X, 400X, and 1000X were used to measure the indentations made over the entire load range for each of the materials. The numerical apertures of the objective lenses corresponding to these magnifications were 0.40, 0.65, and 0.90, respectively. In addition, a stage micrometer was used to verify the accuracy of the measuring system at all magnifications used. Knoop indentation diagonal lengths could be measured repeatedly to within 0.5 $\mu$m.

The following equation was used to calculate the Knoop hardness according to ASTM E384[17] and C 1326[12]:

$$HK = \frac{14229F}{d^2} \qquad (1)$$

where,

HK = Knoop hardness (Kg/mm$^2$)
F = indentation load (g)
d = length of long diagonal ($\mu$m)

The Knoop hardness values obtained from Eq. 1 were converted to GPa by multiplying by $9.80665 \times 10^{-3}$.

SRM RESULTS

The results of indenting the NIST SRM 2830 with the Zwick hardness tester is summarized in Table II. All errors shown are ±1 sample standard deviation (SSD). The Knoop hardness value was in excellent agreement with the NIST value, with less than a 1% difference. Also shown in Table II are the results of the accuracy tests of the digital filar measuring system utilized in the current work. There was less than a 1% difference between the NIST-measured sizes of the NIST-produced indentations and the sizes of the NIST-produced indentations measured in the current study. In addition, measurement of the NIST-produced Knoop indentations in SRM 2830 at both 100x and 400x magnification yielded very similar results, as shown in Table II.

HARDNESS RESULTS

The results of the Knoop hardness testing of the five ceramic materials as a function of load are shown in Table III. In Table IV the corresponding diagonal lengths are shown. Figure 1 graphically represents the hardness as a function of load for the five materials. The following equation was fit to the hardness-indentation load data:

$$HK = \frac{a}{F} + b \qquad (2)$$

where,

a, b = constants
F = indentation load

This equation hereafter shall be referred to as the ISE – load relationship. The first term on the right controls the ISE trend. If a= 0, hardness is independent of load. The coefficients a and b, and the $R^2$ values from the curve fitting, are shown in Table V. Overall, equation 2 did not fit the data well, as seen by the relatively low $R^2$ values in Table V. The AD995 CAP3 material fit the equation the best. Equation 2 predicts that for high loads the hardness should approach the second term, b. This value is also considered the load-independent hardness, or true-hardness, by some people.[11] However, as Figure 1 shows, equation 2 does not fit the hardness data well, particularly the higher load data, at least for these five materials.

The 'a' term in equation 2 gives the magnitude of the change in hardness for a given change in indentation load ($\Delta F$). From Table V it is seen that the SiC materials and the AD995 CAP3 material have similarly high values of 'a' compared to the AlON. The percent hardness change from 0.49 N to 137 N was 25.9%, 38.0%, 35.7%, 32.3%, and 31.3% for the AlON, AD995 CAP3, $\alpha$-SiC, SiC-N, and SiC-B materials, respectively. However, a plot of the 'a' value vs. the percent hardness change did not yield a linear correlation. Again, this means that equation 2 does not fit the data particularly well for these materials.

Table I.  Manufacturer Material Property Data. Data for AlON from Swab et al., Ref 15. Uncertainties are one std. dev.

| | AlON ($\approx Al_{23}O_{27}N_5$) | AD995 CAP3 $Al_2O_3$ | $\alpha$-SiC Hexoloy SA | SiC-N | SiC-B |
|---|---|---|---|---|---|
| Manufacturer | Raytheon | CoorsTek | Carborundum | Cercom | Cercom |
| Processing | proprietary | PS | PS | HP | HP |
| Density (g/cm$^3$) | 3.67 | 3.90 | 3.10 | 3.20 | 3.20 |
| Avg. Grain Size ($\mu$m) | 150 | 6 | 4-10 | 3-5 | 3-5 |
| E (GPa) | 315 | 370 | 410 | 460 | 460 |
| Poisson's Ratio | 0.25 | 0.22 | 0.14 | 0.16 | 0.16 |
| $K_{IC}$ (MPa$\sqrt{m}$) | 2.40 ± 0.11 | 4-5 | 4.6 | 4.7 | 4.4 |
| Knoop Hardness (GPa) | 13.8 ± 0.3 (2 Kg) | 14.1 (1 Kg) | 27.5 (0.1 Kg) | 23.5 (0.3 Kg) | 23.5 (0.3 Kg) |
| Mean 4-pt Flexure Strength (MPa) | 228 | 379 | 380 | 570 | 580 |
| Weibull Modulus | 8.7 | n/a | 8 | 21 | 17 |

Table II.  SRM 2830 Testing Results. (Uncertainties are one std. dev.)

| NIST SRM 2830 HK$_2$ Value (GPa) | Zwick SRM 2830 HK$_2$ Value (GPa) | Percent Difference (%) |
|---|---|---|
| 13.80 ± 0.12 | 13.76 ± 0.06 | 0.7 |
| NIST Measured Knoop Indent Size ($\mu$m) | Current Work Measured Knoop Indent Size ($\mu$m) | Percent Difference (%) |
| 142.0 ± 0.6 | 141.9 ± 0.4 (100x) | 0.1 |
| | 142.3 ± 0.2 (400x) | 0.2 |

Figure 2 shows the variation of the Knoop hardness with the indentation diagonal size. The following equation was fit to the hardness-indentation diagonal length data:

$$HK = \frac{a_1}{d} + a_2{}'$$

(3)

where,
$a_1{}'$, $a_2{}'$ = constants
d = diagonal length

This equation hereafter shall be referred to as the ISE – diagonal size relationship. Again, the first term on the right controls the ISE trend. If $a_1{}'$ = 0, hardness is independent of load.

Table III. Knoop Hardness (Uncertainties are one std. dev.)

| | AD995 | | | | |
|---|---|---|---|---|---|
| Load (N) | AION HK (GPa) | CAP3 HK (GPa) | α-SiC HK (GPa) | SiC-N HK (GPa) | SiC-B HK (GPa) |
| 0.49 | 16.6 ± 1.0 | 20.0 ± 1.5 | 24.1 ± 3.3 | 25.7 ± 2.3 | 24.9 ± 3.2 |
| 0.98 | 16.3 ± 1.0 | 18.4 ± 1.6 | 23.7 ± 2.4 | 23.4 ± 1.2 | 24.0 ± 1.5 |
| 1.96 | 15.5 ± 0.4 | 16.4 ± 1.9 | 24.2 ± 1.0 | 24.2 ± 0.8 | 23.4 ± 0.8 |
| 2.94 | 15.5 ± 0.7 | 16.9 ± 1.0 | 22.6 ± 1.3 | 22.6 ± 0.5 | 22.0 ± 0.7 |
| 4.90 | 14.5 ± 0.7 | 15.4 ± 2.2 | 21.2 ± 0.9 | 21.4 ± 0.5 | 21.0 ± 0.5 |
| 9.81 | 14.1 ± 0.4 | 14.3 ± 0.9 | 19.7 ± 0.9 | 20.2 ± 0.4 | 20.1 ± 0.4 |
| 19.61 | 13.5 ± 0.4 | 13.5 ± 0.5 | 18.1 ± 1.1 | 19.3 ± 0.4 | 19.1 ± 0.2 |
| 50.99 | 12.8 ± 0.4 | 12.9 ± 0.3 | 17.0 ± 0.7 | 17.8 ± 0.2 | 17.5 ± 0.2 |
| 100.03 | 12.4 ± 0.7 | 12.6 ± 0.4 | 16.3 ± 0.6 | 17.7 ± 0.3 | 17.2 ± 0.3 |
| 137.29 | 12.3 ± 0.2 | 12.4 ± 0.3 | 15.5 ± 1.0 | 17.4 ± 0.4 | 17.1 ± 0.1 |

Table IV. Knoop Indentation Diagonal Length (Uncertainties are one std. dev.)

| | AD995 | | | | |
|---|---|---|---|---|---|
| Load (N) | AION (μm) | CAP3 (μm) | α-SiC (μm) | SiC-N (μm) | SiC-B (μm) |
| 0.49 | 20.5 ± 0.6 | 18.7 ± 0.7 | 17.1 ± 1.3 | 16.5 ± 0.8 | 16.8 ± 1.2 |
| 0.98 | 29.3 ± 0.9 | 27.6 ± 1.4 | 24.4 ± 1.3 | 24.4 ± 0.6 | 24.1 ± 0.8 |
| 1.96 | 42.4 ± 0.6 | 41.4 ± 2.5 | 34.0 ± 0.7 | 34.0 ± 0.5 | 34.5 ± 0.6 |
| 2.94 | 52.1 ± 1.2 | 49.8 ± 1.5 | 43.1 ± 1.3 | 43.0 ± 0.5 | 43.7 ± 0.7 |
| 4.90 | 69.4 ± 1.6 | 68.0 ± 7.4 | 57.3 ± 1.2 | 57.1 ± 0.6 | 57.7 ± 0.7 |
| 9.81 | 99.7 ± 1.5 | 99.1 ± 3.2 | 84.3 ± 2.1 | 83.1 ± 0.9 | 83.4 ± 0.8 |
| 19.61 | 143.7 ± 2.2 | 144.1 ± 2.6 | 124.3 ± 3.7 | 120.2 ± 1.2 | 121.0 ± 0.7 |
| 50.99 | 235.6 ± 5.1 | 237.1 ± 2.3 | 206.6 ± 4.1 | 201.9 ± 1.1 | 203.7 ± 1.0 |
| 100.03 | 332.0 ± 3.8 | 336.3 ± 5.4 | 296.1 ± 5.5 | 283.3 ± 2.3 | 287.3 ± 2.4 |
| 137.29 | 396.7 ± 6.0 | 396.7 ± 4.7 | 356.0 ± 11.4 | 334.9 ± 3.6 | 338.1 ± 1.0 |

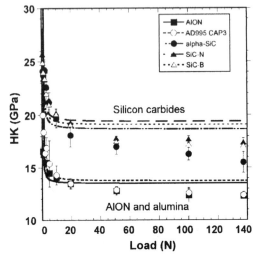

Figure 1. The variation of Knoop hardness as a function of indentation load for the five ceramic materials.

Table V. Hardness Coefficients from HK vs. Load.[*]

| Material | a $(N^2/m^2) \times 10^9$ | b (GPa) | $R^2$ |
|---|---|---|---|
| AlON | 2.0 | 13.5 | 0.65 |
| AD995 CAP3 | 3.6 | 13.7 | 0.81 |
| SiC-N | 3.7 | 19.4 | 0.67 |
| SiC-B | 3.7 | 19.1 | 0.67 |
| α-SiC | 3.7 | 18.6 | 0.51 |

The hardness coefficients from this fit are summarized in Table VI. Overall, the ISE-diagonal size, equation 3, fit the data better than the ISE-load, equation 2, however the fit was still not very good, with the exception of the AD995 CAP3 data, which were fit well. Examination of the hardness data reveals that for all materials the hardness continued to decrease beyond 2 Kg load. The amount of decrease observed, however, varied between the materials. The decreases observed were 1.2 GPa, 1.1 GPa, 2.6 GPa, 1.9 GPa, and 2.0 GPa for the AlON, AD995 CAP3, α-SiC Hexoloy SA, SiC-N, and SiC-B materials, respectively. It is of note that the hardness of the pressureless-sintered silicon carbide (α-SiC) showed a greater decrease compared to the two hot-pressed silicon carbide materials, perhaps because of its large amount of porosity, and that all the SiC materials showed greater decrease compared to the AlON and AD995 CAP3 materials.

---

[*] From Equation 2. Parameters are from a regression analysis and the uncertainties are not yet available.

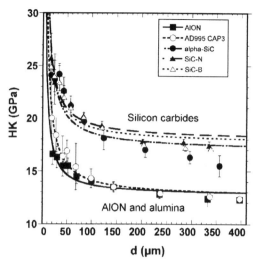

Figure 2. The variation of Knoop hardness with indentation diagonal size for the five ceramic materials.

Table VI. Hardness Coefficients from HK vs. Diagonal Length.[†]

| Material | $a_1'$ (N/m) x10$^4$ | $a_2'$ (GPa) | $R^2$ |
|---|---|---|---|
| AlON | 9.7 | 12.7 | 0.86 |
| AD995 CAP3 | 15.3 | 12.6 | 0.95 |
| SiC-N | 14.6 | 18.0 | 0.87 |
| SiC-B | 14.6 | 17.7 | 0.87 |
| α-SiC | 16.1 | 17.0 | 0.77 |

In addition, the average $HK_2$ value[c] of the α-SiC material was lower by about 1 GPa compared to the hot-pressed SiC's. Presumably, the relatively high volume of porosity (~3 %) and the slightly larger average grain size lead to a lower hardness compared to the hot-pressed materials. Most likely the pores weaken the material, thus allowing the indenter to sink-in further under load. Since pores represent open space in the material, the indenter will have to sink-in further in order to support the load compared to a more fully-dense material. In addition, the slightly larger average grain size of the α-SiC material may lead to a lower hardness as well, since it is known that hardness generally decreases with increasing grain size in the relatively

---

[†] From Equation 3. Parameters are from a regression analysis and the uncertainties are not yet available.
[c] The subscript 2 follows the notation convention of ASTM C 1326 and ISO 14705. The "2" denotes a 19.6 N test load corresponding to a 2 kgf indentation load.

small grain size region.[4,18,19] This phenomenon is usually attributed to Hall-Petch type effects on the plastic deformation, in which grain boundaries tend to block the movement of dislocations[18,20], thus limiting plastic flow and increasing hardness. In addition, the variability of the hardness, characterized by the standard deviation, was greater for the $\alpha$-SiC material compared to the hot-pressed materials. Presumably the wider distribution in grain size (Table I) of the $\alpha$-SiC material would also contribute to the greater variability in hardness. If there is a significant distribution in pore size and/or clustering of pores in the $\alpha$-SiC material, this would presumably further contribute to the variability.

It is noted that for the AlON material, which has an average grain size of ~150 $\mu$m, the hardness values up to and including the 2 Kg load value are essentially single-crystal values, since the average diagonal size was less than the average grain size. This means the ISE appears to be present for single crystals of this material as well. The hardness of the AlON was nearly identical to that of the alumina.

CRACKING AND HARDNESS

It was observed for the three silicon carbide materials that, in general, indentations which were more severely cracked had lower Knoop hardness values compared to less-cracked indentations. This should not be surprising since the more damaged material facilitates greater indentation penetration. From these experiments it could not be ascertained how much of the cracking occurred during load application versus during load withdrawal. Any cracking that occurred during load application affects the diagonal lengths. The indentation tips and their vicinity had negligible cracking and hence, the indentation tips do reflect the maximum extent of indentation penetration while load was applied. (The long diagonals of the Knoop indentations have very little elastic recovery).

An example of the obvious effect of cracking on diagonal length for SiC-N is shown in Figures 3A and 3B. Much less cracking was observed around the mid-section of the indentation in Figure 3A compared to the indentation in Figure 3B. There is a pronounced 14 $\mu$m difference (4.2%) in diagonal lengths due to the cracking. The increased cracking around the indentation in Figure 3B appeared to be primarily in the form of shallow lateral cracking near the surface, however, a generalized increase of damage beneath the surface can not be ruled out either. An example of cracking affecting the hardness of the pressureless-sintered silicon carbide ($\alpha$-SiC) is shown in Figures 4A and 4B. Figure 4C is a higher magnification image of Figure 4B, showing the extensive cracking in the mid-portion of the indentation in greater detail. Although in general it was found that lower hardness could be associated with increased cracking around indentation sites, it should be pointed out that some indentation sites which appeared more-cracked on the surface had the same, or higher, Knoop hardness than less-cracked sites. One reason for this may be that looking at just the surface damage around indentation sites may not accurately assess the full extent of damage in all cases, which exists below the surface as well. That is, the sub-surface damage may not always be accurately reflected in the surface damage surrounding the indentation site, and may be the reason for such discrepancies. Some of the cracking may have occurred during indenter withdrawal as well. Figure 5 is another example of cracking affecting the size of the indentation, and hence the hardness, in $\alpha$-SiC. Quinn et al.[10] showed that cracking either had no effect or decreased the Knoop hardness of glasses depending upon the glass composition.

Figure 6 shows a series of 1.96 N indentations in the AlON material. The indentation in the middle appears un-cracked, however, it had a lower Knoop hardness compared to the

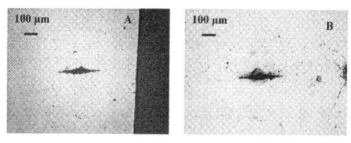

Figure 3. Optical micrographs of 137.3 N (14 Kg) Knoop indentation sites in SiC-N showing variability in cracking. In (A) there is less cracking around the mid-section of the indentation compared to (B). The length of the Knoop long diagonal was 328.6 μm in (A) and 342.3 μm in (B), a 13.7 μm difference, corresponding to Knoop hardness values of 18.1 and 16.7 GPa, respectively. Reflected light, brightfield, nominal 100X mag.

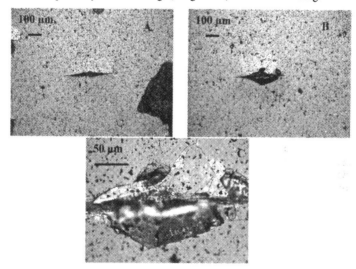

Figure 4. Optical micrographs of 137.3 N (14 Kg) indentation sites in the pressure-less sintered silicon carbide (α-SiC Hexoloy SA). (A) Little cracking around indentation site (d: 362.9 μm, HK: 14.8 GPa). (B) Indentation site with more severe cracking (d: 381.5 μm, HK: 13.4 GPa). (C) Higher magnification image of (B).

indentations above and below, both which exhibited lateral and radial-type cracks. One reason for this may be that since the average indentation sizes at this load (see Table IV) were much less than the average grain size of this material, hardness anisotropy effects due to different crystallographic orientation of the grains with respect to the hardness imprint may have nullified any effects of cracking on hardness. Similar observations were found for other loads as well. It

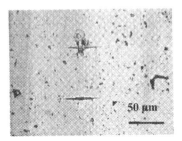

Figure 5. Optical micrograph of 4.90 N indentation sites in the pressure-less sintered silicon carbide ($\alpha$-SiC Hexoloy SA). Top indentation site shows more severe cracking compared to bottom indentation site. Top site (d: 59.0 $\mu$m, HK: 20.0 GPa). Bottom site (d: 57.1 $\mu$m, HK: 21.4 GPa). Reflected light, brightfield, nominal 500X mag.

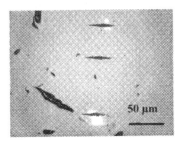

Figure 6. Optical micrograph of 1.96 N Knoop indentations in the AlON material. The middle indentation appears un-cracked at the surface, however it had a slightly lower Knoop hardness compared to the indentations above and below, which were cracked (white area around the mid-point of the indent). Reflected light, brightfield, nominal 500X mag.

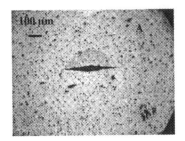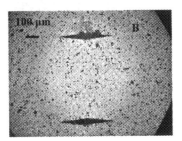

Figure 7. Optical micrographs of 137.29 N indentation sites in the AD995 CAP3 Al$_2$O$_3$ material. (A) Little surface spalling (d: 388.6 $\mu$m, HK: 12.9 GPa) (B) Top indentation: Areas of surface spalling are apparent (d: 402.4 $\mu$m, HK: 12.1 GPa); Bottom indentation: No surface spalling seen, however, indentation is of similar size to top indentation (d: 399.9 $\mu$m, HK: 12.2 GPa).

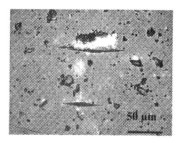

Figure 8. Extreme case of cracking causing a larger indentation (top indentation) compared to a relatively less-cracked indentation (bottom indentation) for the AD995 CAP3 alumina material. Note large chipped region surrounding top indentation. Indentation load was 4.90 N for both indentation sites. Top site (d: 98.2 μm). Bottom site (d: 65.5 μm).

may also be true that the middle indentation landed on the occasional pore in this material, which would account for the lower hardness.

Figure 7A shows a 137.3 N indentation site in the AD995 CAP3 alumina material. Less surface damage exists around this site compared to the top indentation site in Figure 7B. The length of the long diagonal in Figure 7A was 388.6 μm, while the long diagonal of the top indentation in Figure 7B was 402.4 μm. This is a difference of 13.8 μm, but in terms of hardness only a difference of 0.8 GPa, which, however, was larger than the standard deviation at this load by almost a factor of three. The hardness number becomes less sensitive to changes in diagonal length at higher loads, since a given difference in diagonal length is a smaller percentage of the total indentation length as load increases. A most extreme case of cracking affecting hardness in this material is shown in Figure 8. An audible cracking noise was detected during the indentation test that produced the severely cracked indentation site (top indentation in Figure 8).

Additional work in progress is intended to identify the damage processes that account for the ISE and the role of cracking in particular. The repeatability and reproducibility of the ISE trend parameters will also be explored. The correlation of the ISE trends to ballistic impact resistance will be investigated. Larger variability in hardness, whether due solely to variability in cracking or damage around indentation sites, may be an indicator of larger variability in, and possibly poor, ballistic performance.

CONCLUSIONS

The Knoop hardness of five ceramic materials was found to be sensitive to the indentation load used in the range 0.49 N to 137.3 N. In general, a higher Knoop hardness number results when lower indentation loads are used. Cracking around indentation sites in three silicon carbide materials and AD995 CAP3 alumina were found to significantly affect the Knoop hardness. Despite the cracking, the indentation tips were intact and the diagonal sizes readily measured. Cracking usually lowered the apparent hardness and the more cracking, the more the hardness decreased. In addition, it appears that Knoop hardness of ceramic materials continues to decrease beyond 19.6 N (2 Kg) load, although not as sharply compared to in the lower load region. Since cracking is an unavoidable and inherent response of these materials to sharp indentation, it must be viewed as another contribution to the ISE and part of the overall variability of the hardness of brittle ceramic materials. This necessitates making accurate hardness measurements in order to detect subtle differences in variability between different materials which is not due to operator or equipment differences. The overall poor fit of both the ISE – load and ISE – diagonal size equations to the hardness data, particularly at the higher loads where more deviation was apparent, may be due to the effects of some unseen new type of cracking or damage below the surface which is not present at the lower loads. It may also be related to a step change in hardness at some critical cracking threshold.

## ACKNOWLEDGEMENTS

The authors thank the Center for Ceramic Research at Rutgers, The State University of New Jersey, and J.J. Swab for providing the specimens used in this study.

## REFERENCES

[1]J.B. Quinn and G.D. Quinn, "Indentation Brittleness of Ceramics: A Fresh Approach," *J. Mater. Sci.,* **32** 4331-4346 (1997).

[2]J.J. Swab, "Recommendations for Determining the Hardness of Armor Ceramics," *Int. J. Appl. Ceram. Technol.,* **1** [3] 219-225 (2004).

[3]G.D. Quinn, N.D. Corbin, and J.W. McCauley, "Thermomechanical Properties of Aluminum Oxynitride Spinel," *Am. Ceram. Soc. Bull.,* **63** [5] 723-725, 729-730 (1984)

[4]A. Krell, "A New Look at the Influences of Load, Grain Size, and Grain Boundaries on the Room Temperature Hardness of Ceramics," *International Journal of Refractory Metals & Hard Materials,* **16** 331-335 (1998).

[5]R. Berriche and R.T. Holt, "Effect of Load on the Hardness of Hot Isostatically Pressed Silicon Nitride," *J. Am. Ceram. Soc.,* **76** [6] 1602-1604 (1993).

[6]Z. Li, A. Ghosh, A.S. Kobayashi, and R.C. Bradt, "Indentation Fracture Toughness of Sintered Silicon Carbide in the Palmqvist Crack Regime," *J. Am. Ceram. Soc.,* **72** [6] 904-911 (1989).

[7]D.R. Tate, "A Comparison of Microhardness Indentation Tests," *Transactions of American Society for Metals,"* **35** 374-389 (1945).

[8]F. H. Vitovec, "Stress and Load Dependence of Microindentation Hardness," pp. 175-185 in *Microindentation Techniques in Materials Science and Engineering, ASTM STP 889.* Edited by P.J. Blau and B.R. Lawn. American Society for Testing and Materials, Philadelphia, 1986.

[9]K. Hirao and M. Tomozowa, "Microhardness of $SiO_2$ Glass in Various Environments," *J. Am. Ceram. Soc.,* **70** [7] 497-502 (1987).

[10]G.D. Quinn, Patrice Green, and Kang Xu, "Cracking and the Indentation Size Effect for Knoop Hardness of Glasses," *J. Am. Ceram. Soc.,* **86** [3] 441-448 (2003).

[11]H. Li and R.C. Bradt, "The Indentation Load/Size Effect and the Measurement of the Hardness of Vitreous Silica," *J. Non-Cryst. Sol.,* **146** [ ] 197-212 (1992).

[12]ASTM Standard C 1326-99, Standard Test Method for Knoop Indentation Hardness of Advanced Ceramics, ASTM *Annual Book of Standards,* Vol. 15.01, ASTM, West Conshohoken, PA, 2005.

[13]ASTM Standard C 1327-99, Standard Test Method for Vickers Indentation Hardness of Advanced Ceramics, ASTM *Annual Book of Standards,* Vol. 15.01, ASTM West Conshohoken, PA, 2005.

[14]ISO 14705 (2000) "Fine Ceramics (Advanced Ceramics, Advanced Technical Ceramics) – Test Method for Hardness of Monolithic Ceramics at Room Temperature." International Organization for Standards, Geneva, 2000.

[15]J.J. Swab, G.A. Gilde, P.J. Patel, A.A. Wereszczak, and J.W. McCauley, "Fracture Analysis of Transparent Armor Ceramics," pp. 489-508 in *Fractography of Glasses and Ceramic IV.*

Edited by J.R. Varner and G.D. Quinn. The American Ceramic Society, Westerville, OH, 2001.

[16]SRM 2830, Knoop Hardness of Ceramics, Standard Reference Materials Office, NIST, Gaithersburg, MD 20899, 1995.

[17]ASTM Standard E 384-89, "Standard Test Method for Microhardness of Materials," ASTM *Annual Book of Standards*, Vol. 03.01, Philadelphia, PA., 1993.

[18]R. W. Rice, C.C. Wu, and F. Borchelt, "Hardness-Grain-Size Relations in Ceramics," *J. Am. Ceram. Soc.*, **77** [10] 2539-2553 (1994).

[19]A. Krell, "Improved Hardness and Hierarchic Influences on Wear in Submicron Sintered Alumina," *Materials Science and Engineering*, **A209** 156-163 (1996).

[20]R.W. Hertzberg, *Deformation and Fracture Mechanics of Engineering Materials*, 2nd ed. John Wiley & Sons, New York, 1983.

# INFLUENCE OF MICROSTRUCTURE ON THE INDENTATION-INDUCED DAMAGE IN SILICON CARBIDE

Jeffrey J. Swab,* Andrew A. Wereszczak#, Justin Pritchett*, and Kurt Johanns#

*U.S. Army Research Laboratory
Weapons and Materials Research Directorate
Aberdeen Proving Ground, MD 21005

#Oak Ridge National Laboratory
Ceramic Science and Technology Group
Oak Ridge, TN 37831

## ABSTRACT

It is well documented that the microstructure (porosity, grain size, grain boundary composition, etc.) will influence the hardness of a ceramic. However it is not fully understood how these changes may affect the damage that results from the indentation process. In this effort, the density and grain size were varied in silicon carbide (SiC) ceramics. The materials were then subjected to Knoop and spherical indentations. After indentation the resulting damage in each material was assessed, compared, and correlated to the hardness and variations in the microstructure.

## INTRODUCTION

Early work by Lawn and his colleagues[1,2,3] showed that ceramics had a threshold load below which cracking would not develop and at loads below this threshold deformation was essentially plastic. During the development of a new brittleness index Quinn and Quinn[4] conducted extensive hardness testing to develop Vickers hardness-load curves for a variety of brittle ceramics. This index was based on the ratio of deformation energy to fracture energy and they found a distinct transition point between the deformation at low loads and the fracture at higher loads that corresponded to their new brittleness index.

Silicon carbide is one of the leading ceramic materials for a variety of armor applications. One of the primary reasons for this interest is that it is an inherently hard material and high hardness has long been considered a measure of good ballistic protection. The hardness of single crystal[5,6], sintered[4] and hot pressed[7,8] SiC materials is well documented in the literature. While hardness can be considered a measure of a material's resistance to deformation, it is not an inherent physical property of the material. To better understand the deformation resistance of SiC, information of the type and extent of damage generated during ceramic/indenter interactions would be beneficial.

Previous studies of Vickers indentation-induced damage in SiC by Lankford and Davidson[9,10] showed that a plastic zone with a depth about five times the indent diagonal length develops beneath the indentation. Additional examinations[11,12,13,14,15] of the damage created in SiC revealed cleavage as the dominate deformation mechanism in single crystal and polycrystalline SiC with dislocation and stacking faults serving as preferred nucleation sites for twinning and/or cleavage cracks. Hot-pressed SiC subjected to dynamic impact experiments exhibited transgranular fracture with dislocations and stacking faults observed in the material.

This work also reported that there was a substantial increase in the 6H polytype and a decrease in the 3C and 15R polytypes on the surface of the fracture specimens.[7,15] Work by the lead author of this paper examined the damage generated on the surface and beneath Knoop indentations in an armor grade SiC. The analysis supported these earlier finding as cleavage and intergranular cracking were found inside the indentation, while a small damage zone develops beneath the indent with a median crack that forms at the bottom of the damage zone and propagates into the bulk material along the grain boundaries. This paper will summarize an effort to examine the effects of changing grain size and porosity on the damage generated during the indentation process.

EXPERIMENTAL PROCEDURE

Two commercially available SiC materials were obtained from Cercom Inc[*] to examine the effect of grain coarsening on hardness and damage. One was coarse-grained (SiC-N) which is the armor-grade version and the other was fine-grained (SiC-SC-1R) which is for electronic application. Both of these materials were subjected to five hour heat treatments at 2250°C and 2500°C. These heat treatments increased the grain size of the SiC-N from ≈ 4μm in the as-received state to approximately 9 and 10μm respectively, and the SiC-SC-1R from ≈ 1μm to approximately 5 and 7μm. Additional details on these heat treatments schedules can be found in Reference 16. An older vintage of the SiC-N ceramic with three different densities was also examined to determine the effect of density/porosity on the resultant damage. The three different densities examined were 100%, 96% and 90% with the lower density of the latter two versions achieved by decreasing the hot pressing pressure during fabrication. Because this SiC-N was an older vintage (mid 1990s) the fully dense material was included in the analysis and compared to the more recent version.

Specimens from each of the above materials were mounted in epoxy and polished to a quarter micron finish for indentation testing and examination of the respective microstructures. Pointed indentations were made with a Knoop indenter with a load of 19.6N using a Tukon 300[†] microhardness tester, based on the earlier work of the lead author[8], and following the procedure outlined in ASTM C1326[17]. Spherical indentations were made using a 300μm diameter diamond tip attached to the crosshead of a universal test machine. Indentations were made at a rate of ≈ 15N/s, up to 120N with the load recorded as a function of time for both the load and unloading cycle. An acoustic sensor set to a minimum level of 45dB was used to capture the onset of cracking and damage. The damage generated on the surface of the SiC using both indenter geometries was analyzed using optical and electron microscopy.

RESULTS AND DISCUSSION

*Knoop Hardness* - Knoop hardness/load curves were generated for each of these materials to determine the extent of any hardness change due to the variations in the microstructures. These curves are shown in Figure 1 and as expected, the hardness decreases with increasing grain size and decreasing density. Rice nicely summarized the general trend of decreasing hardness with increasing grain size for a variety of ceramic materials[18] and it has been shown that lower density in a sintered SiC results in a decrease in hardness and the brittleness[4]. The HK/load curves for the heat treated SiCs is in excellent agreement with hardness data previously published[16]. An earlier study by the lead author showed that the function – HK =

---

[*] BAE Systems Advanced Ceramics (formerly Cercom Inc), Vista CA
[†] Wilson Instruments, a division of the Instron Corporation, Canton, MA

$(a_1/P) + a_2$ (where HK is the Knoop hardness, P is the indentation load, and $a_1$ and $a_2$ are constants) - adequately described the change in hardness as a function of load for a variety of armor ceramics and that the Knoop hardness was essentially load independent at 19.6N and higher.[8] The hardness data reported here indicates that this might not be entirely true for SiC. Most of the curves in Figure 1 show that the hardness at 98N and in some cases even at 49N, is well below the value predicted by the curve. Attempts to determine the reason for this continued decrease in hardness with increasing load is beyond the scope of this paper. However, it may be due to the extent of fracture and possibly even the type of fracture and cracking that is occurring at these higher indentation loads. In this paper the damage created by a Knoop indenter with an indentation load of 19.6N will be analyzed.

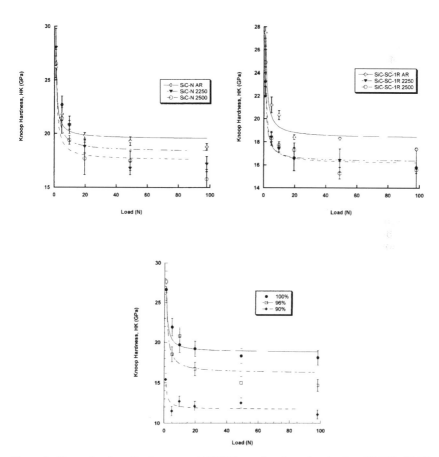

Figure 1. Knoop hardness/load curves. A) SiC-N as a function of grain size; B) SiC-SC-1R as a function of grain size; C) SiC-N as a function of density/porosity.

*Knoop Indentation Damage* – The Knoop indentation damage generated in the as-received and heat treated SiC-N is shown in the top pair of images in Figure 2. In the as-received material the mid-point of the indent is dominated by intergranular fracture and grain boundary cracking. Away from the mid-point towards the indent tips there is evidence of plastic deformation. This is very similar to the damage that results from 9.8N Knoop indentations in the same material.[15] The subsequent heat treatments not only increased the grain size and decreased hardness but also shifted the deformation mechanisms. These indentations, the middle and bottom pairs of images in Figure 2, do not exhibit the extensive intergranular and grain boundary cracking at the mid-point. Instead there is more deformation in the form of cleavage cracks with some grain boundary cracks propagating from the sides of the indent. While the intergranular fracture is not evident on the surface in and around the indent this type of cracking may be present just underneath the surface.

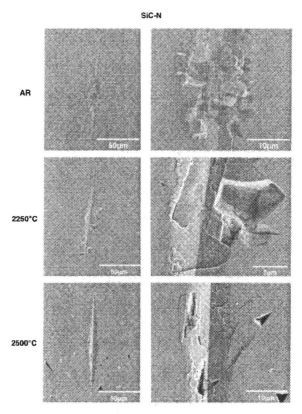

Figure 2. Examples of the damage created by a Knoop indenter with a 19.6N load in SiC-N heat treated to increase the grain size. Image on the left shows the overall indent while the images on the right are high magnification image taken at the mid point of the indent.

Unlike the SiC-N, the as-received, fine-grained SC-1R material exhibited intergranular cracking and fracture along the entire length of the indent, top pair of images Figure 3. Occasionally however, a Knoop indentation would encounter a very large grain. Whenever a large grain was indented the damage was no longer dominated by fracture but shifted to deformation in the form of cleavage cracks, see top pair of images in Figure 3. Indentations into the heat treated versions of this material did not result in the same intergranular fracture. Instead the indents appear cleaner, with some cleavage and grain boundary cracking in the middle of the indent as well as some evidence of crushing. The crushing is most likely due to the presence of porosity that developed during the heat treatment process.

**SiC-SC-1R**

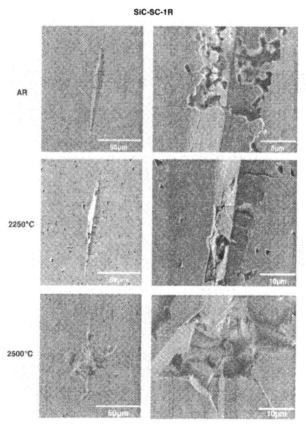

Figure 3. Examples of the damage created by a Knoop indenter with a 19.6N load in SiC-SC-1R heat treated to increase the grain size. Image on the left shows the overall indent while the right is a high magnification image taken at the mid point of the indent.

The damage in the older vintage, but fully dense, SiC-N was very similar to the damage observed in the current SiC-N vintage. This was not necessarily surprising since the hardness values and trends shown in Figure 1 are very similar for both vintages. However, an optical examination of the polished surfaces of both vinatages showed that there is significantly more free C present in the older vintage. It appears that the presence of this free carbon had minimal affect on the hardness or the material's response to the indentation process.

As expected, the decrease in density resulted in a decrease in the amount of fracture associated with the indent but a corresponding increase in the amount of crushing that was present. This is especially true in the 90% dense material where crushing is the dominant deformation mechanism, bottom pair of images Figure 4.

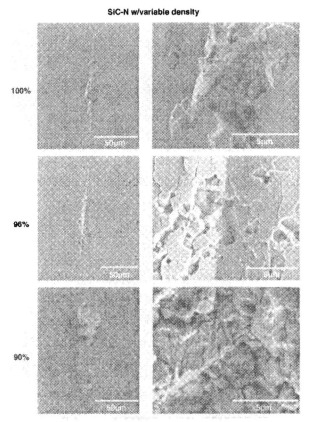

Figure 4. Examples of the damage created by a Knoop indenter under a 19.6N load in SiC-N with different densities. Image on the left shows the overall indent while the right is a high magnification image taken at the mid point of the indent.

*Spherical Indentation Damage* – Reported here are the results from the spherical indentations of the SiC-N with varying density level. The spherical indentation testing of the SiC-N and SiC-SC-1R materials that were heat treated are discussed in reference 19. The fully dense version exhibited plastic deformation inside the indent with a ring of damage approximately 50μm wide around the entire indent, top pair of photos in Figure 5. This damage was probably due to grain boundary and cleavage cracking. Radial cracks initiated from this ring and radiated about 100μm away from the indent along the grain boundaries. The indents in the 96% dense material also had deformation inside and a ring of damage ringing the indent but this ring was not as wide as for the ring in the 100% dense materials, middle pair of photos in Figure 5. Additionally the radial cracks that developed where much shorter. Crushing was the dominant damage mechanism observed in the 90% dense material, bottom photos in Figure 5.

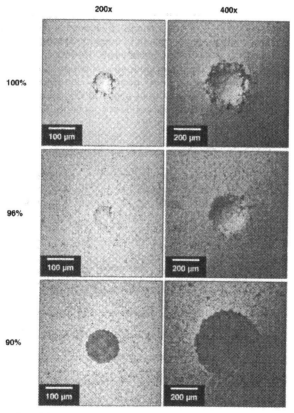

Figure 5. Examples of the spherical indentations and associated damage in SiC-N with varying density.

These differences were confirmed by the acoustic emission data that was captured during the loading and unloading cycles. The acoustic emission data for the 100% was flat indicating that none of the damage events emitted a cumulative energy of 45dB or higher while the 96% material had a single instance early in the loading cycle that exceeded this level. The 90% dense material on the other hand had numerous acoustic events above the 45dB threshold that are most likely due to the extensive crushing that was observed.

SUMMARY

This paper summarized an on-going effort to identify and characterize the damage that develops during the indentation process. The results presented here show that for these SiC materials increasing grain size resulted in a change in the damage mechanisms. Fracture and intergranular cracking tended to dominate at the fine grain size, but with increasing grain size damage in the form of cleavage cracks could be observed. Decreasing the density resulted in a complete shift in damage from fracture and intergranular cracking in the fully dense material to crushing in a SiC that had a density approximately 10% lower. This information can be used to better understand how ceramics respond to an impact event. However, additional information on the damage generated underneath these indentations as well as information on the properties of the grain boundary phases present would greatly expand the understanding of the damage that develops. Finally, while these microstructural variations clearly influence the HK values and lead to a shift in deformation mechanisms, these changes are not sufficient to alter the general trend of the HK/load dependence.

ACKNOWLEDGEMENTS

The research of J. Pritchett was supported in part by an appointment to the Research Participation Program at the U.S. Army Research Laboratory (ARL) administered by the Oak Ridge Institute for Science and Education through an interagency agreement between the U.S. Department of Energy and ARL.

REFERENCES

[1] B.R. Lawn, and D.B. Marshall, "Hardness, Toughness, and Brittleness," *J. Am. Ceram. Soc.,* 62 [7-8] 347-350 (1979)
[2] B.R. Lawn, T. Jensen, and A. Arora, "Brittleness as an Indentation Size Effect," *J. Mat. Sci. Lett.,* 11, 573-575 (1976)
[3] B.R. Lawn, A.G. Evans, and D.B. Marshall, "Elastic/Plastic Indentation Damage in Ceramics: The Median/Radial Crack System," *J. Am. Ceram. Soc.,* 63 [9-10] 574-581 (1980)
[4] J.B. Quinn and G.D. Quinn, "Indentation Brittleness of Ceramics: A Fresh Approach," *J. Mat. Sci.,* 32, 4331-4336 (1997)
[5] O.O Adewoye and T.F. Page, "Anisotropic Behaviour of Etched Hardness Indentations," *J. Mat. Sci. Lett.,* 11, 981-984 (1976)
[6] G.R. Sawyer, P.M. Sargent, and T.F. Page, "Microhardness Anisotropy of Silicon Carbide," *J. Mat. Sci.,* 15, 1001-1013 (1980)
[7] C.J. Shih, M.A. Meyers, V.F. Nesterenko, and S.J. Chen, "Damage Evolution in Dynamic Deformation of Silicon Carbide," *Acta. Mater.,* 48, 2399-2420 (2000)
[8] J.J. Swab, "Recommendations for Determining the Hardness of Armor Ceramics," *Int. J. Appl. Ceram. Technol.,* [3] 219-225 (2004)

These differences were confirmed by the acoustic emission data that was captured during the loading and unloading cycles. The acoustic emission data for the 100% was flat indicating that none of the damage events emitted a cumulative energy of 45dB or higher while the 96% material had a single instance early in the loading cycle that exceeded this level. The 90% dense material on the other hand had numerous acoustic events above the 45dB threshold that are most likely due to the extensive crushing that was observed.

## SUMMARY

This paper summarized an on-going effort to identify and characterize the damage that develops during the indentation process. The results presented here show that for these SiC materials increasing grain size resulted in a change in the damage mechanisms. Fracture and intergranular cracking tended to dominate at the fine grain size, but with increasing grain size damage in the form of cleavage cracks could be observed. Decreasing the density resulted in a complete shift in damage from fracture and intergranular cracking in the fully dense material to crushing in a SiC that had a density approximately 10% lower. This information can be used to better understand how ceramics respond to an impact event. However, additional information on the damage generated underneath these indentations as well as information on the properties of the grain boundary phases present would greatly expand the understanding of the damage that develops. Finally, while these microstructural variations clearly influence the HK values and lead to a shift in deformation mechanisms, these changes are not sufficient to alter the general trend of the HK/load dependence.

## ACKNOWLEDGEMENTS

The research of J. Pritchett was supported in part by an appointment to the Research Participation Program at the U.S. Army Research Laboratory (ARL) administered by the Oak Ridge Institute for Science and Education through an interagency agreement between the U.S. Department of Energy and ARL.

## REFERENCES

[1] B.R. Lawn, and D.B. Marshall, "Hardness, Toughness, and Brittleness," *J. Am. Ceram. Soc.*, 62 [7-8] 347-350 (1979)

[2] B.R. Lawn, T. Jensen, and A. Arora, "Brittleness as an Indentation Size Effect," *J. Mat. Sci. Lett.*, 11, 573-575 (1976)

[3] B.R. Lawn, A.G. Evans, and D.B. Marshall, "Elastic/Plastic Indentation Damage in Ceramics: The Median/Radial Crack System," *J. Am. Ceram. Soc.*, 63 [9-10] 574-581 (1980)

[4] J.B. Quinn and G.D. Quinn, "Indentation Brittleness of Ceramics: A Fresh Approach," *J. Mat. Sci.*, 32, 4331-4336 (1997)

[5] O.O Adewoye and T.F. Page, "Anisotropic Behaviour of Etched Hardness Indentations," *J. Mat. Sci. Lett.*, 11, 981-984 (1976)

[6] G.R. Sawyer, P.M. Sargent, and T.F. Page, "Microhardness Anisotropy of Silicon Carbide," *J. Mat. Sci.*, 15, 1001-1013 (1980)

[7] C.J. Shih, M.A. Meyers, V.F. Nesterenko, and S.J. Chen, "Damage Evolution in Dynamic Deformation of Silicon Carbide," *Acta. Mater.*, 48, 2399-2420 (2000)

[8] J.J. Swab, "Recommendations for Determining the Hardness of Armor Ceramics," *Int. J. Appl. Ceram. Technol.*, [3] 219-225 (2004)

# Author Index

Author Index